NON-STABLE STARS

INTERNATIONAL ASTRONOMICAL UNION

SYMPOSIUM No. 3

HELD IN DUBLIN, 1 SEPTEMBER 1955

NON-STABLE STARS

EDITED BY

GEORGE H. HERBIG

Lick Observatory, University of California

PRINTED WITH FINANCIAL ASSISTANCE
FROM U.N.E.S.C.O.

CAMBRIDGE

AT THE UNIVERSITY PRESS

1957

CAMBRIDGE
UNIVERSITY PRESS

University Printing House, Cambridge CB2 8BS, United Kingdom

Cambridge University Press is part of the University of Cambridge.

It furthers the University's mission by disseminating knowledge in the pursuit of education, learning and research at the highest international levels of excellence.

www.cambridge.org
Information on this title: www.cambridge.org/9781107593602

First published 1957
First paperback edition 2015

A catalogue record for this publication is available from the British Library

ISBN 978-1-107-59360-2 Paperback

CONTENTS

vi

PREFACE

The Executive Committee of the International Astronomical Union, at a meeting in Liège in July 1954, decided to hold a Symposium on the subject of Non-Stable Stars during the Ninth General Assembly of the Union in Dublin, Eire, in August–September 1955. The organization of this Symposium was placed in the hands of a committee consisting of V. A. Ambartsumian (chairman), P. Swings, and G. H. Herbig.

Clearly, if interpreted in a general sense, the subject of instability in stars would be far too large for adequate treatment in the time available for this Symposium. The organizing committee therefore agreed that the subject should arbitrarily be limited to certain areas that are of particular interest at the present time, or that seem to deserve special attention. Invitations to speakers were consequently issued on this basis. Unfortunately, this decision had the effect of eliminating some important topics—such as that of the cepheid variables—from all but incidental discussion. The papers presented at the Symposium, as its organization emerged following considerable correspondence between members of the committee and with other astronomers, fell in these general categories: instability in low-luminosity stars of both late and early type, in the hot stars of rather high luminosity, in late-type variable stars, and in close binary systems. The final section of the Symposium dealt with certain theoretical aspects of stellar instability.

At the Symposium, on account of the pressure of time, only condensed versions of the longer papers could be presented, but their full texts are contained in the present volume. The Soviet astronomers must be commended for having distributed at the Symposium a booklet containing the full versions, in Russian and in either English, German, or French, of most of the papers that were read by members of their delegation. The present volume includes the English texts of these same papers, in a few cases with additions or emendations supplied by their authors in Dublin. It includes also several Soviet contributions that were not circulated at Dublin.

The discussions that followed each section of the Symposium are reported here in a very condensed form. In many cases, the participants in these discussions kindly gave the secretary of the Symposium a written outline of their remarks, but where this was not done, and the sense of the corresponding discussion has not been reported faithfully, imperfect note-taking by the undersigned is alone responsible.

Acknowledgement must be made of the co-operation that the participants in the Symposium afforded those who were responsible for seeing this volume into and through the Press. Finally, the friendly advice and efficient assistance furnished during the pre- and post-Dublin periods of the Symposium by Prof. O. Struve and Prof. P. Th. Oosterhoff, President and General Secretary of the International Astronomical Union, respectively, are most gratefully acknowledged by the organizing committee as a group, and as an individual, by

GEORGE H. HERBIG

LICK OBSERVATORY
MOUNT HAMILTON, CALIFORNIA

18 June 1956

I. INSTABILITY IN DWARF STARS OF LATER TYPE: T TAURI STARS AND RELATED OBJECTS, AND THE UV CETI VARIABLES

1. ON THE NATURE AND ORIGIN OF
THE T TAURI STARS

GEORGE H. HERBIG
Lick Observatory, Mount Hamilton, California, U.S.A.

It is fitting that, in opening this Symposium with a consideration of the T Tauri stars, we note that this month marks the tenth anniversary of the September 1945 issue of the *Astrophysical Journal*, in which appeared the remarkable pioneering paper by Alfred H. Joy that initiated the study of emission-line stars associated with nebulosity. This contribution opened a new approach to the study of the relationship of stars to their environment. Today, ten years later, the T Tauri stars and their interaction with nebular material form a topic whose significance we may not fully appreciate and whose opportunities have been as yet only superficially exploited.

In the decade that has passed since the appearance of Joy's 1945 paper, a great deal of work has been done in this field (most of it, I should remark, by astronomers who are to participate in this Symposium). Rather than attempting to summarize these ten years of progress, I wish to do only two things: first, to present an assessment of the problem of the nature of the T Tauri stars as it stands today, and second, to discuss some evidence that may bear on the question of the origin of these objects. If a satisfactory solution to even one of these problems can be found, it will then be possible to look for a deeper significance in the instability of the T Tauri stars.

For those who are not familiar with the subject, it should be said that the T Tauri stars are low luminosity objects having a rather characteristic emission-line spectrum, are irregularly variable in light and spectrum, and have been found only in association with nebular material, both bright and dark. The observational data are now so extensive that the reality of the association with nebulae is quite beyond dispute. The central question is: are the T Tauri variables new stars, recently formed or still forming within the nebulae, or are they ordinary field stars that have encountered the gas and dust clouds accidentally, and are in process of being modified—perhaps rather superficially—by their environment?

Three years ago [1], in an examination of the information then available, I concluded that it was 'not possible to make a clear-cut and entirely acceptable decision between [these] two opposing alternatives on the basis of observational evidence alone'. Today, in my opinion, we are able to take a more positive stand. I believe that the evidence now available favours the hypothesis that the T Tauri stars as a class are new objects, genetically associated with the clouds in which they are found. Presumably their instability is a consequence of their relative youth. Some of the evidence to this effect is still ambiguous, but I believe that if one were to maintain the opposite position, and try to explain away all the observational facts by *ad hoc* processes involving normal stars, he would find himself in an unrealistic position.

First consider three observational facts that must be mentioned here because of their undoubted significance, but which in my opinion do not by themselves enable us to reach a firm decision.

(1) It has been found[1] that the T Tauri stars have somewhat diffuse absorption lines, a characteristic which is known to be quite abnormal for ordinary single stars of the same spectral types[2]. Although only the brightest T Tauri stars are accessible to observations of this kind, there is indication that other stars of the group also may possess this peculiarity.

(2) The T Tauri stars of later type are found to be systematically brighter than main sequence stars of the same spectral types[1]. Some of the M-type objects in the Taurus clouds, for example, lie 2 to 3 magnitudes above the main sequence even before any allowance is made for absorption. Similar departures from the main sequence have been found in NGC 2264[3].

(3) If the T Tauri stars are field stars in a stage of active accretion, one might expect that their spectra would give some evidence of the infall of material. However, there is no clear spectroscopic evidence that matter is actually being collected by the T Tauri stars. In fact, the conclusion which emerges from both low-[4] and high-dispersion[5] spectroscopic observations is that the emission-line regions of their atmospheres, as well as certain higher layers that produce a shell-type spectrum, are *rising* with respect to the source of the underlying absorption spectra. Since the absorption spectra presumably indicate the radial motion of the stars, one would expect to observe just such an effect if the T Tauri stars were ejecting material, rather than collecting it.

These three characteristics—line width due to rapid rotation or large-scale turbulence, abnormal luminosity, and lack of evidence for the infall of matter—neither demonstrate nor disprove that the T Tauri stars are newly-formed objects. It is true that newly-formed stars might exhibit abnormally high rotation as a result of the conservation of angular momentum during formation, and that the ejection of material would be a way of disposing of excess momentum. It is also true that the abnormally

4

bright T Tauri stars inhabit a portion of the spectrum-luminosity diagram where lie the evolutionary tracks of gravitationally contracting stars having a mass somewhat less than the sun's [6, 7]. But as interesting as it would be to identify the phenomena observed in the T Tauri stars with these theoretical concepts, one cannot use such correspondences to prove the correctness of the basic assumption. It might be that the collection of a great deal of material by a normal star could somehow give rise to these same effects. Fortunately, there are three other points which bear on this problem that cannot be evaded so readily.

(4) The data that are now available on the occurrence of emission-line stars in the Taurus dark clouds [4, 8, 9] make it possible to compute, under reasonable assumptions, the average number of such stars per cubic parsec of these nebulae. In the central region of the Taurus clouds, it is found that there are, depending upon the initial assumptions, from 5 to 15 times as many T Tauri stars per unit volume, down to about apparent visual magnitude 17·5, as there are stars in the same absolute magnitude interval in the region of the sun. Allowance for undetected T Tauri-like variable stars which (temporarily or permanently) show no emission lines might increase these figures from 50 to 100 %. In other words, there are far more T Tauri stars in these clouds—by one order of magnitude—than can be accounted for by the random encounter of field stars with the clouds.

(5) In some nebulae, there have been found curious concentrations of emission-line objects in the vicinity of a much brighter early-type star imbedded in the cloud. In this small region, the star density may become rather large. The most obvious explanation of such a concentration is that, for some reason, stars are actually being formed there, since field stars would be scattered at random through the nebula. Whether the presence of the early-type star is the cause, or is a consequence of the favourable conditions for star formation prevailing in that area is an open question.

If, then, the T Tauri stars are newly-formed objects, it might be asked if there is any observational indication as to how they begin their careers. This is a dangerous question to attempt to answer at the present time because of the considerable variety of objects that are found in association with nebulae. Our lack of any general understanding of the processes which are taking place may well cause one to attach undue importance to phenomena which are, in reality, not of major significance. Furthermore, one must refrain from generalizing too far on the basis of limited information, because it is entirely possible that the T Tauri stars can originate in more than one way. With these reservations, let us consider some observational results which appear to bear upon this question.

5

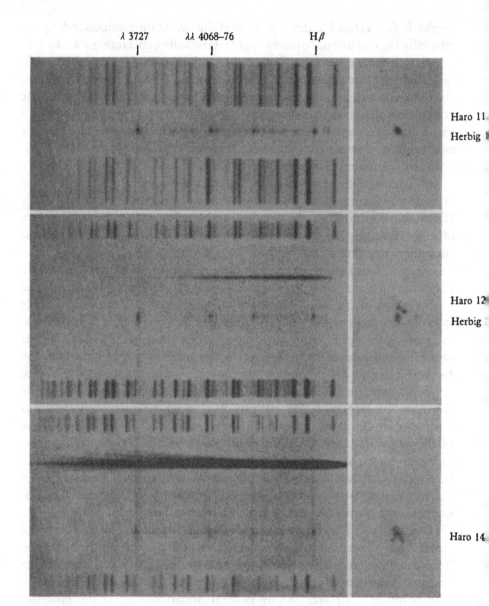

λ 3727 $\lambda\lambda$ 4068–76   Hβ

Haro 11
Herbig

Haro 12
Herbig

Haro 14

Fig. 1. Slit spectrograms and direct photographs of three Herbig-Haro Objects in Orion. The dispersion of the original spectrograms is 430 Å/mm. at Hγ; they were obtained with the nebular spectrograph of the Crossley reflector of the Lick Observatory. The direct photographs, which do not match the spectrograms either in scale or in orientation, were obtained with the same telescope and with blue-sensitive plates. In all the direct photographs, north is at the top and east to the right. The spectrogram of Haro 12a refers to the knot of stars and nebulosity at the lower (southern) edge of the Object.

In a dense dark cloud not far from the Orion Nebula lie the brightest known examples of a very peculiar class of nebulous stars that have been called the 'Objects of Group D' by Herbig[1], and the 'Herbig-Haro Objects' by V. Ambartsumian[10]. (I hope that I shall be forgiven if I employ here the latter designation, which seems less clumsy than the other.) They have been studied by Herbig[11], Haro[12], and, very recently, by K.-H. Böhm[13]. Very briefly, the simplest of these Objects consist of a stellar or semi-stellar nucleus together with a very small emission nebula, whose strongest lines beside the Balmer series are those of [O I], [O II], and [S II], which occur with rather unusual relative intensities (see Fig. 1).

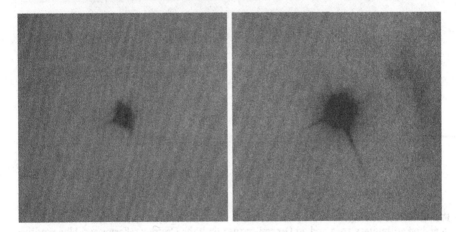

Fig. 2. The nebulosity immediately surrounding T Tauri. Both photographs were obtained with the 100-inch Mount Wilson reflector by Dr W. Baade, to whom I am indebted for the privilege of reproducing them here. North is below and east to the left. The left-hand photograph is a short exposure obtained on 19 December 1944, and the right-hand is a longer exposure taken on 1 October 1940. The latter photograph shows, on the extreme right, the brighter portion of NGC 1555, Hind's Variable Nebula. The elongation of the image of T Tauri by bright nebulosity in position angles 151° and about 331°, discovered visually by Burnham, is evident, as is a nebulous flare in about 190°. The sharp bright rays in p.a. 60°, 150°, etc....are due to the supports of the Newtonian flat mirror of the telescope.

Accompanying this nebular spectrum, at least in the brightest Objects, is a weak, low-temperature continuum and a set of bright lines due to easily produced ions, such as Ca II and [Fe II]. This second spectrum, although heavily masked by the nebular emission lines, is reminiscent of the spectra of some T Tauri stars.

Consequently, it is a fact of extraordinary interest that T Tauri itself is surrounded by a very small nebula[14] (see Fig. 2) whose bright-line spectrum is very similar to those of the Herbig-Haro nebulae. If the source of the T Tauri-like spectrum in the brightest Herbig-Haro Objects—

7

presumably a star—should simply brighten up by 4 or 5 magnitudes, then the result might be a rather good match for T Tauri and its nebulosity.

These results suggest the possibility that at least some of the stars like T Tauri may have begun their histories as Herbig-Haro Objects[10]. Speculations that these Objects are somehow connected with star formation have been encouraged by the recent discovery that two objects which seem to be stellar have appeared in one of the brighter Herbig-Haro Objects sometime between January 1947 and December 1954 (see Fig. 3).

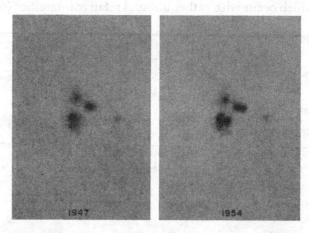

Fig. 3. Direct photographs of the Object Haro 12a=Herbig 2, in Orion. The plates were obtained on 20 January 1947 and 20 December 1954 with the Crossley reflector and blue-sensitive emulsions. North is at the top and east at the right. The scale can be obtained from the fact that the two brightest stars on the 1947 photograph are 8″ apart.

The 1946–7 Lick photographs show this particular Object to have a complex structure: it is composed of a pair of stars 8″ apart of about photographic magnitude 17, probably 3 much fainter stars, and several small masses of nebulosity, all in an area about 20″ across. On the 1954–5 plates, two additional stars are present, each lying 3 to 4″ from a component of the original pair of 8″ separation. Photographically, these newly-appeared objects have brightened up through ranges of at least 3 magnitudes. Their photographic magnitudes are now about 17, but much of this energy is concentrated in the emission lines. Direct photographs in the wave-length region $\lambda\lambda\,5200$–5800, which is essentially free of bright lines, show that in this spectral region the stars as well as the nebulosity are at least one magnitude fainter than in the photographic region. The 'visual' absolute magnitudes, with the contribution of emission lines excluded but without allowance for selective absorption, must therefore be about $+9$ at the present time. No large changes have been observed in the brightness of these newly-appeared objects during two months of the 1954–5 observing season, but one would be rash to attempt a prediction of their behaviour in the years to come.*

Our understanding of what may be taking place in this Object could hardly be more incomplete, but perhaps we have witnessed the opening phase of an episode in stellar evolution. It may be in this way that a T Tauri star begins its history, and I have emphasized that possibility here. But the structure of the particular Herbig-Haro Object in which these two 'new' stars have been found suggests an alternative evolutionary interpretation, as follows: We know that the T Tauri stars often occur in pairs, and occasionally in loose groups, but this Object now contains seven star-like objects within a very small area. The space density is about 10^5 stars per cubic parsec. Such a concentration is highly reminiscent of the complex multiple star systems often found in O-associations [15], of which a well-known example, the Trapezium of the Orion Nebula, lies $1°5$ away. Therefore, perhaps this particular Herbig-Haro Object is near the beginning of the evolutionary road that leads upward toward the high-luminosity O- or B-type multiple systems. The question of whether the T Tauri stars are to be identified with an intermediate stage along this road, or whether they define an entirely separate path, forms a subject for future investigation.

One way in which one might learn more of the relationship of the T Tauri stars to the Herbig-Haro Objects is to find stars in an intermediate state of development, under the assumption that they would be recognizable as such. A search for objects of this type has been underway at the Lick Observatory for some time, and one promising prospect has been found.

(6) It is now possible to make an additional point regarding the nature of the T Tauri stars, which it was necessary to postpone until now. The Lick direct photographs show that the newly-appeared objects in Orion certainly did not develop out of any stars that had been visible at those positions before. If these 'new' stars are to be identified with the beginning of the development of a T Tauri star, then this evidence (within its limitations as to absolute magnitude) also supports the hypothesis that the T Tauri stars are not rejuvenated interlopers from the surrounding field, but are, in fact, newly-formed objects. The same conclusion is to be drawn, of course, if the newly-appeared objects are the beginnings of O- or B-type stars.

* Plates taken in September and December 1955 showed that no further change had taken place in the appearance of this Object since the observations of December 1954–February 1955.

REFERENCES

[1] G. H. Herbig, *Jour. R.A.S. Canada*, **46**, 222 (1952).
[2] G. H. Herbig and J. F. Spalding, Jr, *Ap.J.* **121**, 118 (1955); *Contr. Lick Obs.* Ser. II, No. 56.
[3] M. F. Walker, *Sky and Telescope*, **13**, 426 (1954); *Ap. J. Suppl.* **2** (No. 23), 365 (1956).
[4] A. H. Joy, *Ap.J.* **102**, 168 (1945); **110**, 424 (1949).
[5] R. F. Sanford, *Publ. A.S.P.* **59**, 134 (1947); G. H. Herbig, unpublished.
[6] E. E. Salpeter, *Mem. Soc. Roy. Sci. Liège*, quatr. série, **14**, 116 (1954).
[7] L. G. Henyey *et al.*, *Publ. A.S.P.* **67**, 154 (1955).
[8] G. Haro, B. Iriarte and E. Chavira, *Bol. Obs. Tonantzintla y Tacubaya*, No. 8, 8 (1953).
[9] G. H. Herbig, unpublished.
[10] V. Ambartsumian, *Comm. Burakan Obs.* No. 13 (1954).
[11] G. H. Herbig, *Ap.J.* **113**, 697 (1951).
[12] G. Haro, *Ap.J.* **115**, 572 (1952); **117**, 73 (1953).
[13] K.-H. Böhm, *Publ. A.S.P.* **67**, 338 (1955); *Ap. J.* **123**, 379 (1956).
[14] G. H. Herbig, *Ap.J.* **111**, 11 (1949).
[15] S. Sharpless, *Ap.J.* **119**, 334 (1954).

2. IRREGULAR VARIABLE STARS OF THE RW AURIGAE TYPE

P. N. KHOLOPOV

Astronomical Council, Academy of Sciences, Moscow, U.S.S.R.

The RW Aurigae type was first introduced by P. P. Parenago[1] in 1932 for the designation of irregular variable stars that belong to spectral class G and show extremely rapid (up to 1 magnitude in several hours) and large (up to 3 or 4 magnitudes) variations of their brightness. Many stars of that type were found to be connected with dark diffuse nebulae. Some of them, as for instance R Monocerotis, RY Tauri, and R Coronae Australis, were also associated with bright and dark comet- or fan-shaped nebulae, often themselves variable.

There existed a double terminology for the designation of the irregular variables connected with the diffuse nebulae. Such stars are often named the Orion Nebula type variables, because a particularly numerous group of irregular variables that consists, according to Parenago, of not less than 220 members[2] is connected with this Nebula. The term 'RW Aurigae type variables' also is used for the designation of isolated variable stars of the above-mentioned properties. In the course of time it has frequently happened that the same stars have been referred by some observers to the RW Aurigae type, and by others to the Orion Nebula type, owing to their striking connexion with the dark diffuse matter.

The paper by Joy on 'T Tauri Variable Stars' appeared in 1945[3]. The following features should be present, according to Joy, in order to assign a star to this new type: (a) irregular variations of brightness with an amplitude of about 3 magnitudes, (b) an F5 to G5 type spectrum with Hα and Ca II emissions, as in the spectrum of the solar chromosphere, (c) low luminosity, and (d) connexion with a dark or bright nebula.

Joy revealed, in fact, one of the most important properties of the RW Aurigae type variables: the existence of a typical emission spectrum, which had escaped the attention of earlier observers. At the same time, he introduced the term 'T Tauri type variable', which soon became popular along with the designation 'RW Aurigae type'.

In 1949 Hoffmeister, in his paper on 'RW Aurigae Type Stars and their Related Types',[4] described the results of his studies of the light variations of these variables. The chief criterion for referring stars to the type, which Hoffmeister proposed to call the RW Aurigae type in an extended sense, is solely, in his opinion, the general outline of the light curve. Hoffmeister believes that an essential feature of this type of variable is the existence of rapid, non-periodic fluctuations, interrupted by intervals of rest (the novae and U Geminorum type stars are excluded). The replacement of intervals of activity by periods of weak or slow variation, or even of constant brightness, characterizes the RW Aurigae type stars. In general, variations of 0·5 to 1 magnitude per day, or sometimes per hour, are considered to be rapid. Such variations are usually not observed in the case of slow irregular variables which are red giants. Rapid variations are often superimposed upon slow variations of the mean brightness.

The term 'in an extended sense' refers to the fact that among the variables resembling RW Aurigae in the character of their light variation, stars are met not only of class G, but also of all the other spectral classes from B to M. The distances of the dark nebulae, in the directions of which the majority of variables of that type are concentrated, and the spectral properties established for some of these stars, testify that the absolute magnitude of the variables approximate those of main sequence stars of corresponding spectral classes.

In 1946 Joy[5] discovered forty faint stars having the Hα line bright that were located in the regions of dark nebulae in Taurus, Auriga, and Orion. Their emission spectra appeared to be extremely similar to the spectra of T Tauri type variables, described by him earlier; they were mainly dwarfs of spectral classes A–M[6].

In 1951 it was discovered by me, together with N. E. Kurochkin, that almost all of these stars were also variables that belonged, from their spectral properties, to the RW Aurigae type[7].

At about the same time clusterings of stars having bright Hα were discovered in a number of other diffuse nebulae by Struve and Rudkjøbing[8], Haro and Moreno[9], and also by Herbig[10]. Many of these stars were found to be variables. Emission lines were also found in the spectra of a number of irregular variables in the nebulae of Orion[11, 12] and NGC 2264[13]. Further study of the spectra of these variables[14, 15] showed their similarity to the spectra of the RW Aurigae type variables that are connected with the dark clouds in Taurus and Auriga.

Thus, there are no grounds at present to retain the term 'stars of the type of the Orion Nebula variables' along with the term 'RW Aurigae type variables', in so far as there is no essential difference between these two notions.

Several attempts to classify the RW Aurigae type variables have been made. Hoffmeister's classification [4] is based only upon the outline of the light curve. This is the reason for its restriction. The classification by Parenago [16] is, to some extent, similar to Hoffmeister's. It was suggested by Parenago in the course of his study of irregular variables in the Orion Nebula, and is based upon the frequency distribution of magnitudes of a given variable. It may be supposed, however, that the distribution of magnitudes can only be used to describe the behaviour of an irregular variable in a given epoch, but not for its classification.

At the same time, among the most typical representatives of the T Tauri type (as, for instance, T Tauri and RW Aurigae), variables with both rapid and slow variations of their brightness are met. It may, therefore, be believed that the rate of light variation does not characterize the type, but rather the state of the star, and perhaps also its age.

The most advisable classification—according to physical properties— has not yet been developed. There are grounds for its creation, however, that are afforded by the studies of Joy and Herbig, who suggested schemes for the classification of the emission stars connected with the diffuse nebulae.

Similarly to Hoffmeister, who introduced the RW Aurigae type in an extended sense, Herbig has practically introduced the T Tauri type in an extended sense, as well. He assigned to that type not only the Ge stars, but also stars of much later types, up to M, inclusively. The difference between the two systems is that Herbig bases the type upon the spectral features— in particular on the emission spectrum and not upon the photometric characteristics.

In the case of Hoffmeister's classification, the spectral class and luminosity are not essential data. In Herbig's classification, contrary-wise, the character of light variations is not essential, in so far as it is assumed that the majority of (or perhaps all) stars with emission spectra of the T Tauri type are variable.

The question about the correspondence of the two types arises immediately. Do all the stars referred to the RW Aurigae type according to their photometric features possess spectral characteristics inherent to the T Tauri type in Herbig's sense, and vice versa? Such a question can at once be answered negatively. Thus, for instance, R Monocerotis has a spectrum typical of the T Tauri type stars, but has never shown any rapid light variations and is not referred by Hoffmeister to the RW Aurigae type. T Tauri itself is a similar case. We refer these stars to the RW Aurigae type on the basis of their spectral features. On the other hand, Herbig

13

showed[17] that the majority of variables that have been referred to the RW Aurigae type on the basis of their photometric behaviour alone, and which are not, as a rule, members of systems associated with diffuse nebulae, do not have emission spectra.

Herbig concludes that the RW Aurigae type variables in Hoffmeister's sense do not form a physically homogeneous group. He doubts even the physical reality of this type. His conclusion is based on the fact that the spectral types of a number of RW Aurigae type stars are contained in the limits from B5 to M4, while their luminosity classes range from Ib (super-giants) to V (dwarfs). In Herbig's opinion, it is not easy to find a common cause of the light variations that could act in objects as different as are the RW Aurigae type stars in Hoffmeister's sense.

However, such a cause could be processes arising from the non-stability of these stars. It is possible that all these objects, regardless of their spectral classes and luminosities, are of recent origin and that their luminosities are not yet quite stabilized.

It is known that a number of RW Aurigae type stars in the Orion and NGC 2264 nebulae have no bright lines in their spectra. On the other hand, as was established by Herbig and Haro, the intensity of Hα emission is often variable for such stars and may completely disappear in their spectra.

Thus the RW Aurigae type consists both of emission and non-emission variables, of extremely diverse physical properties. The question of the homogeneity of the type and the similarity or difference of the causes of light variation of the stars related to that type is very complicated, however, and various possibilities must be taken into account in its solution.

In connexion with the actual variety of the physical properties of the RW Aurigae type variables it is extremely important to have a criterion that permits one to judge with some certainty whether a star belongs to that type. This criterion can be sought in the fact that the cause of the light variability of the examined stars is the same.

The main argument supporting a unique cause of the light variation of the majority of RW Aurigae type variables lies in the fact that these variables are, as a rule, found together, in groups which we now call T-associations. This term was first introduced by V. A. Ambartsumian[18], who noticed that of eleven bright T Tauri type variables studied by Joy, seven are located in the same area of the sky, in the region of Taurus and Auriga. Some groups of the RW Aurigae type variables as, for instance, in the Orion Nebula, around R Coronae Australis, and elsewhere, were recognized a long time ago. At present, twenty-three T-associations are

14

known to us [19]. It must be noted that T-associations are found in the form of nuclei in O-associations. They are known for about all of the nearest O-associations. At the same time, doubtlessly, there exist T-associations that do not enter into the composition of O-associations and do not contain O or Bo type stars.

What is the cause of the light variation of the RW Aurigae type stars? For several decades, it was considered that the RW Aurigae type stars are 'obscured' variables. Later on, when it was found that the light variations of many such stars are accompanied by changes of their spectral classes, it became clear that the cause of the light variations should be sought in physical processes taking place in the upper layers of the stars. The Orion Nebula variables, which were for a long time the classical examples of the obscured variables, had to be excluded from that category.

This does not mean, of course, that obscured variables do not exist at all. No convincing example may be given at the present time, however, where the light variation is caused by a change of the optical thickness of the diffuse matter lying between the star and the observer.

If the change of the conditions of obscuration plays a part in the light variations of the above-mentioned emission variables, this part is not an essential one. It should, however, be emphasized that there are no indisputable proofs of the assumption that the nature of the non-emission and emission stars of the RW Aurigae type is similar. The assumption of their similarity is supported by the variability of the emission spectra and the similarity of their distributions in the nebulae. Strictly speaking, the arguments against the hypothesis of obscuration are based upon the fact that a quite improbable structure of the dark nebula is required in order to explain the observed rapid light variations of these stars. That is, extremely large local density gradients or small condensations are demanded. But if it were assumed that T-associations are regions of star formation, then it cannot be denied that the structure and properties of the nebulae connected with T-associations might really be of an extraordinary character. Although it seems to us that the non-emission variables represent the subsequent stage of the evolution of emission variables, the role of obscuration in the light variation of the non-emission RW Aurigae type stars cannot be altogether rejected at present.

The variability of the emission spectrum accompanying the light variation of the emission stars cannot, of course, be explained by obscuration. It is caused by some other more complicated physical processes.

There are at present two points of view regarding the nature of these processes. It is considered, on the one hand, that the RW Aurigae type variables are ordinary stars which entered the dark cloud at random and are changing their brightnesses owing to processes of interaction with the surrounding diffuse matter. It is sufficient to eliminate, from this point of view, the dark cloud, and the emission lines in the spectrum of the star will disappear, and the changes of brightness will stop. On the other hand, it is supposed that the cause of variability lies in the stars themselves, in which case these stars are not ordinary ones.

Investigators of the spectra point out the difficulties in the spectral classification of these objects, the absorption lines often being veiled by strong continuous emission. This emission continues far into the violet. The distribution of energy in the continuum is essentially different from that in the continuous spectrum of an ordinary cool dwarf.

In order to explain the appearance of the emission lines in the spectra of stars connected with nebulae and the light variation of such stars, Greenstein [20] advanced a hypothesis that the cause of such phenomena is the same as in the case of flares in the solar chromosphere according to Giovanelli's theory. In that theory, the electrons are accelerated in the variable magnetic field of sunspots, and atoms are excited through electron collisions. According to Greenstein's opinion, the streams of ionized dark matter falling upon the star create variable magnetic fields of such intensity that the spectrum of the flares becomes visible in the general spectrum of the star.

We admit that hypotheses similar to that of Greenstein may possibly explain the appearance of a faint emission spectrum in the case of ordinary dwarfs which enter the dark cloud at random. Such emission may become visible if the processes of the interaction of the star with the surrounding diffuse matter become sufficiently important. We do not know, however, the extent to which these effects are appreciable. It should be extremely important to establish with certainty the spectroscopic effects caused by such interaction, in order to be able to separate them from the properties inherent to the spectrum of the star itself.

Haro, Iriarte and Chavira [21] called attention to the fact that stars showing strong Ca II, but weak Hα emission, and belonging to the CO Orionis association, are located in the H II region embracing the system of λ Orionis. In the region of dark nebulae in Taurus and Auriga, where there is, apparently, much neutral hydrogen, bright Hα lines in the spectra of emission stars are extremely intense. Haro, Iriarte and Chavira believe that this indicates that the spectral composition of radiation of the emission star depends upon the state and nature of the surrounding diffuse matter. We do not see, however, any causal relationship between the above phenomena.

Returning to the hypothesis by Greenstein, we have to point out that it is extremely difficult to admit that the light variation of from 1 to 3 magnitudes observed in RW Aurigae type variables might be explained by means of this hypothesis. It is doubtful if a source of emission not connected with the star might, under its influence, emit an amount of energy exceeding the energy of radiation of the star itself by two or three, or even ten times.

It must be taken into account that the production of the phenomena observed in the spectra of the RW Aurigae type stars can take place without the participation of diffuse matter. This is proven by the fact that in the spectra of UV Ceti type stars (variable dwarfs located in the vicinity of the Sun, with spectral classes from dM3e to dM6e) similar emissions are observed during the outbursts, and Hα and Ca II emission lines are observed in them permanently, between the flares.

Open clusters containing G-type stars (as, for instance, the Pleiades), with which dust clouds are connected, are known. The density of these clouds can hardly be less than the density of the peripheral parts of the nebula in which such a typical emission and variable star as RW Aurigae itself is located. In the Pleiades, the stars of which represent excellent photometric standards, no faint stars similar to RW Aurigae are found. They should have been discovered a long time ago if they were present.

This permits one to suggest that the emission and variability of the RW Aurigae type stars is caused not by the diffuse matter, but by the stars themselves.

It is of great interest to recall the hypothesis suggested by V. A. Ambartsumian [22] and I. M. Gordon [23], to the effect that the continuous emission observed in the spectra of the RW Aurigae type variables and UV Ceti type stars during their outbursts is not of a thermal character. These authors explain this phenomenon in terms of the radiation emitted by relativistic electrons in the hypothetical magnetic fields of these stars. Ambartsumian suggested also another explanation of that phenomenon, involving the direct release of energy by stellar matter from the deep interior that is ejected into the external atmospheric layers of these stars.

RW Aurigae type stars cannot be ordinary stars that have entered the diffuse nebulae at random, as otherwise the fact of their concentration into small volumes of comparatively large diffuse nebulae cannot be understood. Greenstein and Herbig showed that the spatial density of stars in these nebulae exceeds considerably the density of ordinary stars of corresponding types in the vicinity of the Sun.

T-associations are actual clusters. Their formation cannot be explained by the collisions of dwarf stars with nebulae, and the gradual accumulation of the stars therein by a process of capture by the nebulae. This mechanism has been described by L. E. Gurevitch[24]. The spectral composition of different T-associations, some of which contain mainly one type of stars while others contain stars of a different kind, would otherwise be inexplicable. Thus, for example, there are no G-type stars in the CO Orionis association, but many K_3–K_4-type stars. In the UZ Tauri association, at the same time, a number of G-type stars are found, but none of types K_3–K_4[7].

Neither can it be assumed in such cases that a whole cluster of ordinary dwarfs has entered a nebula, because some explanation would then have to be given of the fact that the cluster is found exactly in the centre of the diffuse nebula, as is observed in the Orion Nebula and in NGC 2264. The genetical connexion of these systems of stars with the surrounding nebulae has been suggested by a number of investigators.

The average density of some T-associations is so low that they can hardly be considered dynamically stable systems[7].

In a number of T-associations, nuclei (regions of increased density) are observed. Such nuclei are striking in all of the more or less studied T-associations: those in Taurus, Auriga, Orion and Monoceros. The richest T-association in Orion has four nuclei, located along a straight line with equal intervals between their centres, thus forming a chain. This is the system of Orion's Sword, which contains the famous Trapezium in one of its nuclei. Such a system can by no means be considered a random formation, or an old one. It may be of recent origin, having had as yet no time to disintegrate.

The following facts also testify that stars belonging to the T-associations are unusual.

Herbig[13] discovered that the absorption lines of four typical emission stars (T Tauri, UX Tauri, RY Tauri and SU Aurigae) are anomalously wide and diffuse. This is never observed in the case of ordinary dwarfs of the corresponding spectral classes. According to Herbig's opinion, the phenomenon can be explained as due to either high rotational velocities of these dwarfs, or to strong turbulence in their atmospheres. The choice between these two possibilities cannot yet be made, but in either case this phenomenon deserves much attention.

Let us point out also that Sanford[25] has observed that the spectrum of T Tauri suggests the probability of ejection of matter from the atmosphere of that star.

The sum of all the above-mentioned facts testifies that the RW Aurigae type stars are, chiefly, stars of recent origin. This point of view was first proposed by V. A. Ambartsumian [26]. It is not excluded that among the RW Aurigae type variables are to be found ordinary stars which have entered the dark clouds at random. We are unable as yet to separate such stars from emission stars of constant brightness that actually belong to T-associations. It is not impossible that among the non-emission RW Aurigae type variables, so classified according to their photometric behaviour only, there will be found a large number of stars that are erroneously referred to that type. But all this can hardly affect the conclusions that follow from the study of the properties of the majority of variables of that type, which do belong to T-associations.

Herbig restricts the T Tauri type to stars in the vicinity of the main sequence, with spectral classes of type G and later. But there are no reasons for us to believe that the variability of stars of earlier classes and even of different luminosities (for example, giants or sub-giants) that belong to the T-associations differs by its nature from that of stars of later types.

We assign at present to the RW Aurigae type stars those irregular variables which are located on the spectrum-luminosity diagram in the region of the complete main sequence (from spectral classes O to M), and in the region of the sub-giants. These variables possess emission spectra in the majority of cases and are mainly characterized by rapid variations of brightness. However, among them stars are found in whose spectra emission is present only weakly, sometimes disappearing completely, as well as stars having comparatively slow light variations.

By the T Tauri stars, we understand merely a group of RW Aurigae type stars of spectral class Ge, as was initially suggested by Joy.

It is not excluded, that the luminosities of some typical variables of the T Tauri class, like R Monocerotis and R Coronae Australis (i.e. variables connected with comet-like nebulae), will appear to be significantly higher than the luminosities of main sequence stars of corresponding spectral classes. Some of these variables may even be super-giants. This cannot, however, be a reason to exclude such variables from the category under consideration.

The question regarding the direct connexion of UV Ceti type variables with the RW Aurigae type variables remains open for the present time. Even in the nearest T-associations we do not know stars of spectral classes later than dM2·5. The earliest of all the known UV Ceti type stars belong to spectral class dM3e. The so-called rapid variables, which were discovered by Haro and Morgan [27] in the Orion Nebula and which are reminiscent of UV Ceti type stars in the character of their light variation, have higher luminosities and are, obviously, a variety of the RW Aurigae type variables.

These conclusions permit one to formulate the following working hypothesis for the systematization and interpretation of the results of observation. It may be believed that RW Aurigae type variables are stars of recent origin, which are in process of formation (Ambartsumian proposed the use of the term 'stars in the making' for them), and that they define one of the initial stages of the life of main sequence stars, of subgiants, and possibly also of other types.

This does not mean, however, that stars originate initially in the form of RW Aurigae type variables. Ambartsumian [22] indicated that possibly the form that preceeds the RW Aurigae type stars might be the Herbig-Haro Objects: i.e. the small bright round nebulae with which only very faint blue stars can be associated. Only seven objects of that kind are known in the Orion Nebula region, three of them forming a chain of about 5' in length.

We think that RW Aurigae type variables might suddenly originate from a preceding stage. The remarkable star FU Orionis that flared up in the centre of a small dark globule, was found to be connected with a bright fan-shaped nebula. In this case, we actually observed the appearance of the comet-shaped nebula and the star connected with it. It is not excluded that in FU Orionis we are eye-witnesses of the actual formation of an object similar to R Monocerotis or R Coronae Australis, and it is also not impossible that these objects originated like FU Orionis. This star deserves careful watching. It is not impossible that the typical emission spectrum will appear in the future, and slow variations of brightness, similar to that observed in T Tauri and R Monocerotis, will set in. If this hypothesis (being as yet on the verge of fantasy) is correct, then it may be supposed that the rapid light variations set in during the latest stages of the evolution of the RW Aurigae type variables, previous to their transition into a stable state. This conclusion is based upon the fact that slowly varying RW Aurigae type variables appear to be strongly connected with diffuse nebulae, whereas stars of that type whose connexion with the diffuse nebulae is less evident, show mainly rapid light variations.

REFERENCES

[1] P. P. Parenago, *Variable Stars*, **4**, 222 (1933).
[2] P. P. Parenago, *Variable Stars*, **9**, 89 (1953).
[3] A. H. Joy, *Ap.J.* **102**, 168 (1945).
[4] C. Hoffmeister, *A.N.* **278**, 24 (1949).
[5] A. H. Joy, *Publ. A.S.P.* **58**, 244 (1946).
[6] A. H. Joy, *Ap.J.* **110**, 424 (1949).
[7] P. N. Kholopov, *Variable Stars*, **8**, 83 (1951).
[8] O. Struve, M. Rudkjøbing, *Ap.J.* **109**, 92 (1949).
[9] G. Haro and A. Moreno, *Bol. Obs. Tonantzintla y Tacubaya*, No. 7, 11 (1953).
[10] G. H. Herbig, *Publ. A.S.P.* **62**, 142 (1950).
[11] G. Haro, *A.J.* **55**, 72 (1950).
[12] G. H. Herbig, *Ap.J.* **111**, 15 (1950).
[13] G. H. Herbig, *Jour. R.A.S. Canada*, **46**, 222 (1952).
[14] G. Haro, *Ap.J.* **117**, 73 (1953).
[15] G. H. Herbig, *Ap.J.* **119**, 483 (1954).
[16] P. P. Parenago, *Publ. Sternberg Astronomical Institute*, **25**, 225 (1954).
[17] G. H. Herbig, *Trans. I.A.U.* **8**, 805 (1954).
[18] V. A. Ambartsumian, *Bull. de l'Acad. de Sci. de l'U.R.S.S., sér. phys.* **14**, 15 (1950).
[19] P. N. Kholopov, *Transactions of the Fourth Conference on Problems of Cosmogony*, Moscow, 1955.
[20] J. L. Greenstein, *Publ. A.S.P.* **62**, 156 (1950).
[21] G. Haro, B. Iriarte and E. Chavira, *Bol. Obs. Tonantzintla y Tacubaya*, No. 8, 3 (1953).
[22] V. A. Ambartsumian, *Comm. Burakan Obs.* No. 13 (1954).
[23] I. M. Gordon, *Comptes rendus de l'Acad. de Sci. de l'U.R.S.S., nouvelle sér.* **97**, 621 (1954).
[24] L. E. Gurevitch, *Transactions of the Second Conference on Problems of Cosmogony*, p. 235, Moscow, 1953.
[25] R. F. Sanford, *Publ. A.S.P.* **59**, 134 (1947).
[26] V. A. Ambartsumian, *Evolution of Stars and Astrophysics*, Erevan, 1947.
[27] G. Haro and W. W. Morgan, *Ap.J.* **118**, 16 (1953).

3. ON RW AURIGAE TYPE STARS AND RELATED TYPES

C. HOFFMEISTER

Sonneberg Observatory, Sonneberg 15 b, Thüringen, Germany

First of all it will be necessary to say a few words on terminology. Different names are now in use either for the same group of variables or for different variants of a rather heterogeneous class. Let me list the following:

RW Aurigae stars,   Orion type variables,

RR Tauri stars,   Nova-like variables,

T Tauri stars,   Main sequence variables.

Nebular type variables,

The first three names, using the designation of a prototype star, have been formed in analogy to other classes of variables such as the Mira, Algol, RV Tauri, δ Cephei, and RR Lyrae stars. The name T Tauri stars, however, which was introduced by Joy, points to certain properties of the spectra and therefore does not include all objects which might be assigned to the class on the basis of photometric behaviour. Furthermore, T Tauri is photometrically not characteristic of the class. According to Ludendorff, the star is similar to R Coronae Borealis, the changes generally being rather slow. This conclusion is supported by the light curves given by Esch and Losinsky. At my request, Mr Ahnert has estimated T Tauri on all sky patrol and other plates available at the Sonneberg Observatory, and has found the variations from about 1930 up to the present time to be in agreement with the statements mentioned above, with one exception: in September 1934, the variable showed rapid changes of range about 0·6 magnitude. This fact indicates that T Tauri really belongs to the RW Aurigae class, but is far from being a typical member.

Before we can come to a decision as to the best designation, we must examine this class of variables as a whole.

(1) *Photometric properties.* A typical RW Aurigae star is a variable having rapid non-periodic changes whose range is from 1 to 4 magnitudes. The variations may be either continuous or interrupted by short or long intervals of practically constant light. Quasi-periodic fluctuations are not excluded, but are not an essential feature. This behaviour is typical of the

22

class, but there are many variants that have small ranges, relatively slow variations, Algol-like minima occurring at irregular intervals, or behaviour reminiscent of the U Geminorum stars. T Tauri represents an extreme sub-type that generally has slow variations and that only rarely shows rapid changes. It can be seen that the light curves exhibit a large diversity.

(2) *Spectral properties*. In the 1948 *General Catalogue of Variable Stars*, there are 150 stars classified as of the RW Aurigae or Orion types, including those whose assignment of type is followed by a question mark. For no more than twenty-five of these stars is the spectral type given. The distribution is as follows.

RW and RW? stars		Ori and Ori? stars	
Go to G5	11	B2	1
K5	1	Ao to A2	3
M	1	Fo to F5	2
Pec	1	G5	1
		K	4

On the basis of this distribution scheme, the typical RW Aurigae stars are seen to have a rather small range of spectral class; the typical spectral type is dG5e, in contrast to the Orion stars which have a large range in type. Beside these two groups, there is at least one other, represented by the emission-line variables in the Taurus clouds, the characteristic spectral type of which is dKe, thus indicating an association with the flare stars. But it seems that the distribution of energy in the continua of these stars is not that of a genuine dK-type star, for investigations by Goetz and Wenzel at Sonneberg Observatory and by Haro at the Tonantzintla Observatory reveal an excess of blue light. Probably emission lines are a typical feature of the RW Aurigae and Orion variables, but there is no narrow correlation with the type of light variation, because T Tauri has a typical RW Aurigae-like spectrum without being a typical variable of this type.

(3) *Relation to Interstellar Matter*. An outstanding feature is the association of RW Aurigae stars with bright or dark nebulae, well-known examples being T Tauri and R Monocerotis. On the other hand, there are equally typical members that have no appreciable relationship with nebular matter, as, for example, RW Aurigae. Statistically the situation is contradictory. In an investigation which I made some years ago, I found a good correlation of the locations of the bright RW Aurigae variables with interstellar clouds but a surprisingly poor correlation for the faint ones, especially in the Taurus cloud. (The dKe stars found later are not included in this remark.) The explanation seems to be the following: two distinct maxima of frequency with apparent magnitude are shown by variables of

this region, and if one tentatively assumes an absolute magnitude of $+5$ for all the stars, then two groups at distances of about 80 and of about 400 parsecs are indicated. The members of the first group might be regarded as foreground stars projected on the clouds, while the second group is formed by distant stars visible in the windows between the clouds. This interpretation is not in favour of the existence of a uniform physical association. In my opinion the distribution over the sky of the isolated RW Aurigae variables found up to the present time is primarily a result of the method of search, and therefore is not suitable for the study of the spatial distribution on a large scale. But it must be emphasized that the existence of real associations of main sequence variables and their association with nebular matter is beyond doubt. Indeed they seem to be of a different kind than the pretended Taurus association. I would mention the Orion association, the group around S Monocerotis, the group in Corona Austrina, and the association of dKe-type stars in the western part of the Taurus clouds.

Scrutinizing these results, one gets the impression that the problem is a rather complicated one. There are at least three groups of stars whose relationship is probable although not immediately obvious:

(a) Genuine isolated RW Aurigae stars and members of closely related subtypes,

(b) variables in close association with nebulae,

(c) isolated stars whose relationship to Group (a) is rather highly probable.

It is very difficult to define the limits of this last group, for there are some rather heterogeneous objects that may have to be included, such as type Be variables like γ Cassiopeiae, and stars like V Sagittae and EM Cygni that are now regarded as old novae. The situation is complicated the more since no clear separation is possible between the three groups. The first impression is that Groups (a) and (b) are, although related, nevertheless of a different kind. But entering into the details one finds in Group (b) genuine RW Aurigae stars as well as Orion variables that are isolated from nebulosity, and thus obviously belong to Group (a). In this connexion, SX Phoenicis may be mentioned. It is a variable with an extremely short period and strong variations of the light curve. It also seems to be a main sequence star, and not a normal cluster type variable. Further possible members are the novae of different types, and the U Geminorum stars and related variables, which in the Hertzsprung-Russell diagram have positions below the main sequence.

The most important problem to be solved is that of the separation and

24

relationship of the different groups, especially the isolated RW Aurigae stars and the Orion type variables. The problem involves both their physical condition and their position in space. Are they objects of different properties and cosmic significance, or are they related in the same way as are the RR Lyrae stars in globular clusters and those in the general galactic space? It might be suggested that nebular regions may be the place where such stars develop and from which they originate, so that the isolated stars are to be regarded as emigrated objects. One of the means of separating the groups is the determination of spectral types and luminosity classes. A beginning has been made by the work of Herbig at Lick Observatory.

A general aspect of the problem in its present state is that the observational material is by far too limited. This is true not only for the spectroscopic information, as mentioned above, but also for the photometric data. Very much work is required in the future to provide better information on the light curves of this most complicated group of stars, where every individual seems to represent a sub-type of its own, and where only very long and condensed series of observations are able to reveal the details. During my recent stay in South-West Africa, I availed myself of the excellent climate of that country by observing some RW Aurigae stars of different sub-types, but only collaboration on a large scale can lead to complete success, even for a single star.

Returning to the question of terminology I make the following proposal: the general heading of *main sequence variables* includes all the objects mentioned in this report, and is in agreement with the classification scheme given by Schneller in *Geschichte und Literatur II*, **3**, p.v. The discrimination of sub-classes depends primarily upon photometric properties so long as spectral types are not available to a large extent. So we may follow the present usage of the *General Catalogue*, where the main classes are *Novae*, *U Geminorum stars*, and *RW Aurigae stars*. One small change might be proposed: all stars of related properties not showing the typical characteristics of the RW Aurigae class ought to be named *RW Aurigae-similar*, abbreviated RWs, without the introduction of further details. Only future work will show whether or not well-defined sub-types exist.

4. THE POSSIBLE CONNEXION BETWEEN T TAURI STARS AND UV CETI STARS

GUILLERMO HARO

Tonantzintla and Tacubaya Observatories, México, D.F.

As is well known, very rapid and non-periodic changes in brightness have been discovered in several late-type dwarf stars in recent years. In the vicinity of the sun, within a radius not exceeding 10 parsecs, nine or ten such objects have been found and named 'flare' stars because of their extraordinarily rapid variations. The prototype of these flare stars is UV Ceti. For the purpose of the present discussion, we shall call these objects 'classical' flare stars.

On the other hand, at the Tonantzintla Observatory twenty objects have been discovered that are associated with the interstellar clouds in Orion[1, 2], Taurus[3], and Monoceros; they show extremely rapid variations in brightness that are similar to those of the classical flare stars. We shall refer to these as 'flash' stars in nebulae, so as to distinguish them, provisionally, from the classical flare stars.

The remarkably similar characteristics of the flash stars in nebulae and the classical flare stars challenge us to decide if it is possible to combine these two groups of rapid variables into one single type. It is evident that this possible amalgamation will not only clarify the obvious relationship of the flash stars in nebulae with the T Tauri type stars but the relationship between the T Tauri stars and the classical flare stars as well.

To clarify the problem let us first inquire whether it is possible to consider the flash stars in nebulae as examples of a more general type of variable star, and, further, let us investigate with care their possible relationship with the classical flare stars.

The criteria that I should like to propose for distinguishing variable stars of the flash type in nebulae comprise four general characteristics and two particular ones. The four general characteristics are:

(1) The exhibition of short-lived and non-periodic outbursts lasting from a few minutes to approximately two hours.

(2) The spectral types range from at least dK6 down to late dM.

(3) With regard to their general shape, the light-curves are 'unique' and are similar in form to the light-curves of novae: the rise to maximum

is extremely rapid; the decline from maximum may be slower, but it is also rapid. Excepting at the times of the short-lived variations, the star remains near its normal minimum magnitude.

(4) Either the appearance or the enhancement of emission lines may be observed during maximum light.

The two particular characteristics are:

(5) During the outburst, as far as can be learned from very limited observational material, the spectral characteristics are not distinguishable from those of some T Tauri stars; in addition to the bright-line spectrum, a strong emission continuum is present.

(6) The quiescent spectrum shows emission lines.

Characteristics (1) and (3) permit us to distinguish, to a certain extent, a flash star from the common T Tauri type variables, including the T Tauri stars with flare-like spectroscopic characteristics observed by Joy[4]. The four general characteristics are common to all the flash stars known so far, whereas the two particular characteristics (5) and (6) are only observed in some of them.

On the basis of the above criteria, it is now possible to examine the flash stars in nebulae and the classical flare stars together, and point out the similarities and differences to be found not only in the variable stars considered individually within their respective group, but also the two groups compared with each other:

(a) All the rapid variable stars of the classical flare type and of the flash type have the four general characteristics in common.

(b) Some of the stars of the classical flare group and of the flash group show, in addition to the four general characteristics, the two particular ones. Examples: UV Ceti (classical flare) and the Orion Nebula star No. 8 (flash star) [1].

(c) Not all stars of the classical flare group show the particular characteristic (5). Examples are: Proxima Centauri[5] and HD 234677, observed by Popper[6]. Similarly, some stars of the flash group fail to exhibit characteristic (5). Examples are: Orion Nebula No. 11 and Orion Hα No. 71 [2, 1]. This fact suggests that each of the two groups of variable stars may be subdivided into rapid variables with and without characteristic (5).

(d) Excepting HD 234677 of type dK6, where no direct evidence on the rapidity of the variation has been obtained, all the known classical flare stars have spectral types later than dM3; on the other hand, in the flash stars we find spectral types from dK6 to dM6.

(e) All the classical flare stars show emission lines in their spectra at minimum; however, not all the flash stars show permanent emission lines at their minima.

It will be noticed from the foregoing exposition that the same general criteria which allow us to distinguish a flash star as such apply equally well to the case of the classical flare stars. Apart from certain differences in the individual characteristics, which are indiscriminately found among the members of both groups (i.e. the presence of the emission continuum: characteristic (5)), there are, nevertheless, as previously pointed out, certain characteristics in some flash stars which have not been observed in the classical flare stars. It would be most important to know how essential these differences can be. They are as follows:

First, although it is true that among the flash stars in nebulae there are spectral types as early as dK6, there are late dM-types as well. For instance, in the Taurus dark clouds, six out of the seven flash variable stars discovered show spectral types later than dM3 and only one has a dMo spectrum [3] (Dr Kuiper has classified this particular star as dK8); the most frequent spectral type of the flash stars in this region is later than dM3. At the same time, in the Orion Nebula the most frequent spectral type for the flash variable stars is between dK6 and dMo. These two examples—of Orion and Taurus—can lead one to believe that it is highly probable that the spectral type frequency of the flash stars may change significantly as a function of the evolutionary state of the stellar aggregate to which they belong.

If in the Taurus region we find out of seven flash stars, six with spectra later than dM3, it is not strange that in the vicinity of the Sun all the classical flare stars known so far show late dM spectral types; this does not necessarily mean that an essential condition for membership in this class of variable stars is to fall within an extremely restricted range of spectral types. The possible inclusion of HD 234677, classified by Popper as dK6, within the group of classical flare stars should eliminate any doubts on this subject.

The second apparent difference refers to the presence of emission lines in the quiescent spectra. Although many of the classical flare stars were discovered without knowing that their spectra were of the dMe type, the data on the most recent discoveries may be seriously affected by observational bias owing to the fact that it has become the fashion to search for variable stars of the classical flare type only among the dMe stars. This may explain why, up to now, no flare stars without emission lines in their normal spectra have been found near the Sun. As regards the flash stars in nebulae, we usually do not find bright lines in their quiescent spectra. If

28

this difference is not the result of an observational selection, I am unable to estimate its importance at this time. (It is to be noted that in the case of the T Tauri type stars, some show permanent emission lines and others do not, but this does not lead us to treat them as two different classes of variables.)

The remarkable similarities in the characteristics of the flash stars and the classical flare stars make it extremely difficult to separate them into two different intrinsic groups and strongly tempt one to consider them as members of a single physical type of variables. However, the obvious implications resulting from such an amalgamation are so far-reaching that great caution should be exercised. Yet an excessive conservatism could lead us to reject or indefinitely postpone the recognition of a phenomenon that may be of great importance in the study of stellar evolution.

Besides the common characteristics that have been listed for the flash stars in nebulae and the classical flare stars, two important observational arguments can be made in favour of the hypothesis that all belong to only one type of variable star:

(1) *The spectrum-rate of variation relation* [3]. If we consider the classical flare stars and the flash stars together, it becomes evident that, on the basis of the observational data now available, there exists in both groups a relation that connects the spectral types to the total duration of the sporadic outbursts: the later the spectral type, the more rapid the variation. It is extremely unlikely that this spectrum-rate of variation relation is the result of a simple and deceiving coincidence, and is not an intrinsic property that reveals, to a certain extent, the operation of the same phenomenon.

(2) *The kinematic properties of the dMe stars.* Jean Delhaye's study [7] of the kinematic properties of the dMe stars in the vicinity of the Sun (out of twelve dMe stars studied by Delhaye, four are classical flare stars) shows there to be a remarkably small velocity dispersion perpendicular to the galactic plane. Therefore, the dMe stars can probably be considered to be young stars, forming a very flat sub-system. Dr Oort, who kindly called my attention to Delhaye's work, suggests the possibility that these dMe stars are associated with interstellar clouds. It is quite plausible that, in the present case, we are observing a fraction of a stellar sub-system having the structural peculiarities of the T-associations as defined by the Soviet astronomers.

The discovery of the flash stars and their peculiar distribution in the Orion Nebula, led us at the Tonantzintla Observatory to believe that these objects belong to the family of the T Tauri stars, and that whenever there exists a T-association there is the possibility of finding, related to it, rapid variable stars of characteristics similar to those of UV Ceti. Our survey of the Taurus dark clouds and our preliminary observations of the nebulous cluster NGC 2264 have strengthened this belief.

All the considerations presented here support—quite forcefully, notwithstanding a reasonable caution—the hypothesis that both the classical flare stars and the flash stars in nebulae belong to the same class of variables and, therefore, have a similar origin and represent parallel evolutionary paths. If this hypothesis is accepted as fundamentally valid, we must recognize that the stars of the UV Ceti type are to be considered as objects related to the T Tauri stars.

It is interesting and perhaps significant to point out again the difference found in the spectral type frequencies of the flare stars (or flash stars) discovered in Orion, Taurus and in the immediate vicinity of the Sun. While in the Orion Nebula the flash stars have late spectral dK types, the majority of known rapid variable stars in Taurus show spectra later than dM3 and only one has a type of dK8 or dMo. Near the Sun, all the known flare stars, with the possible exception of HD 234677, are later than type dM3. These differences lead one to speculate on the possibility that the flare stars belonging to T-associations of various ages disclose, by themselves, the evolutionary stage of the stellar aggregate to which they belong.

REFERENCES

[1] G. Haro and L. R. Terrazas, *Bol. Obs. Tonantzintla y Tacubaya*, No. 10, 3 (1954).
[2] G. Haro, *Bol. Obs. Tonantzintla y Tacubaya*, No. 11, 11 (1954).
[3] G. Haro and E. Chavira, *Bol. Obs. Tonantzintla y Tacubaya*, No. 12, 3 (1955).
[4] A. H. Joy, *Ap.J.* **102**, 168 (1945).
[5] A. D. Thackeray, *M.N.* **110**, 45 (1950).
[6] D. M. Popper, *Publ. A.S.P.* **65**, 278 (1953).
[7] J. Delhaye, *Comptes rendus*, **237**, 294 (1953).

5. FLARE SPECTRA IN DWARF STARS

ALFRED H. JOY

Mount Wilson and Palomar Observatories, Pasadena, California, U.S.A.

Observational evidence of transient flares which produce conspicuous changes in the total brightness and in the spectra of dwarf stars is limited, at present, to a small number of faint late-type stars. The outbursts take place with extraordinary rapidity and the duration is usually only a few minutes or hours. Accompanying spectral changes such as veiled and fuzzy absorption lines, greatly increased intensity of the bright lines of hydrogen and helium, and the appearance of an emission continuum shortward of λ 3750, have been reported.

Flares of this kind are known in three somewhat diverse groups of stars:

(1) stars involved in diffuse nebulosity;

(2) extremely low-luminosity red dwarfs;

(3) SS Cygni stars.

The sudden flares often seen in the sun are probably analogous to the examples of instability observed in stellar atmospheres. Such solar disturbances are, however, relatively minor in extent. While sufficient to excite line emission, they fail to release the large amounts of energy necessary to emit the hot continuous spectrum observed in the three groups of stars.

(1) DWARF STARS INVOLVED IN DIFFUSE DARK CLOUDS OR BRIGHT NEBULOSITY

In general, these may be classed as T Tauri stars. Flare-like effects have been found spectroscopically in about half of the stars for which slit spectrograms have been obtained. The time involved is apparently an hour or considerably more.

Two sub-groups of this class may be recognized:

(a) Stars of type dGe with many emission lines. These rather rare objects have been extensively observed. Without doubt the large variations of light and spectrum are due to some interaction with the surrounding nebulosity.

Strengthened hydrogen and continuous emission in the λ 3600 region,

31

which probably accompany the flare-like activity, are present on Mount Wilson plates of nearly all of these stars. Also, they have been reported in RW Aur[1], in DD Tau[2], in stars of NGC 2264[3], and in other stars in Orion[4].

In the T Tauri stars the variations in total brightness are often large and comparatively slow, so that it should be possible to separate flares of short duration from the complicating large-scale variations of the star.

Further study is needed to decide upon the nature of the ultra-violet continuum. Simultaneous spectroscopic and photometric observations are required in order to correlate the intensity of the bright hydrogen lines and the ultra-violet continuum with the changes in the total light.

(b) Stars of type dK and dM with few emission lines. More than 700 stars belonging to this group are now known, mostly through the observations of Haro and Herbig. They are extremely faint and are concentrated in the dark and bright gas and dust clouds of the Milky Way.

From representative slit spectrograms it seems that they are much alike, having absorption spectra, on the average, of about type dK5. The H and K lines of Ca II and the Balmer lines of hydrogen appear in emission with an occasional showing of weak bright lines of He I, Fe II, and [S II].

In many of the stars the intensity of the emission lines and of total light are known to vary. Some of the slit spectrograms show enhanced continuous emission without absorption lines extending shortward of λ 3700, and Haro has observed this strengthened ultra-violet radiation on direct photographs exposed through selected filters[4]. Neither the origin of this excessive energy output or its duration are known, but it probably results from some flare phenomenon of short duration rather than a widespread thermal outburst. The ultra-violet continuum is usually accompanied by strong Balmer emission with a slow shortward decrement.

On some spectrograms of these stars, the absorption lines are distinctly shallow and fuzzy as a result of filling-in by overlying continuum, but since this veiling does not *always* accompany the ultra-violet continuum it may have a different origin.

(2) LATE-TYPE dMe STARS OF EXTREMELY LOW LUMINOSITY

Sudden outbursts in total light or spectral changes characteristic of flare activity have been recognized in about twenty dMe stars of the lowest luminosity known. The time involved in such outbursts is usually only a few minutes; in some flares marked changes in brightness have been observed to take place within a few seconds of time. Because the occurrence of such flares is entirely unpredictable, and the exposure times for spectrograms relatively long, little is known of the spectral changes taking place during a flare.

A spectrogram of UV Ceti was fortuitously obtained on Mount Wilson in 1948 at the exact time of such an outburst [5]. While this spectrogram may represent an extreme example of a flare, the remarkable changes observed may be assumed to be characteristic of flare spectra in general. The normal dM5e spectrum was completely dominated by a spectrum showing:

(a) a strong continuum corresponding to a temperature higher than 10,000° K.,

(b) strong hydrogen emission lines 2 angstroms in width with a slow shortward decrement,

(c) bright lines of He I and a faint λ 4686 line of He II.

The observation of these outstanding features has not yet been duplicated in UV Ceti or found in other low-luminosity dMe stars, but lesser indications of the same effects occur on Mount Wilson spectrograms of YZ CMi, HD 196982 B, and on a McDonald spectrogram of Wolf 47 (exposed by Bidelman). Strengthened Balmer lines occur on certain Mount Wilson plates of V1216 Sgr, BD + 19° 5116 B, and DO Cep. Changes in the intensity of hydrogen emission lines have also been reported in the dMe stars 20 C 1250 (Luyten), HD 196982 (Luyten), V371 Ori (Wachmann), V645 Cen (Thackeray), and HD 234677 (Popper). Whether these variations in the spectra were accompanied by increases in the total brightnesses of the stars is not known.

(3) SS CYGNI STARS

Of this group, three stars (AE Aqr, SS Cyg, and RU Peg), at minimum light show absorption dG to dKo spectra upon which are superposed the continuous spectra of hot companions having bright lines 20 angstroms in width. The variations in radial velocity indicate binary motion with periods of less than a day.

Mount Wilson observations of AE Aqr [6] and SS Cyg at minimum light show the characteristic flare effects on some spectrograms, and Lenouvel's photo-electric observations [7] of AE Aqr reveal many flares, of a few minutes in duration, in the total light of the system.

The origin and nature of these light and spectrum changes are very uncertain. Simultaneous observations of light and spectrum are needed to separate processes taking place in the two bodies.

Somewhat similar effects are noted in the spectral changes of the three groups of stars, but the underlying cause of the strange and unexpected behaviour, which may be quite different in the three types, invites both observation and analysis.

REFERENCES

[1] G. H. Herbig, *Publ. A.S.P.* **60**, 256 (1948).
[2] O. Struve and P. Swings, *Publ. A.S.P.* **60**, 61 (1948).
[3] G. H. Herbig, *Ap.J.* **119**, 483 (1954).
[4] G. Haro and G. H. Herbig, *Bol. Obs. Tonantzintla y Tacubaya*, No. 12, 33 (1955).
[5] A. H. Joy and M. L. Humason, *Publ. A.S.P.* **61**, 133 (1949).
[6] A. H. Joy, *Ap.J.* **120**, 377 (1954).
[7] F. Lenouvel and M. Golay, *Comptes rendus*, **237**, 1215 (1953).

6. PHOTO-ELECTRIC OBSERVATIONS OF AE AQUARII

F. LENOUVEL

Observatoire de Haute Provence, St-Michel (Basses-Alpes), France*

Dr A. H. Joy has shown that the star AE Aquarii is a spectroscopic binary. Seventy hours of photo-electric observations have revealed light variations accompanied by frequent explosions.

The light curve of AE Aquarii shows a roughly sinusoidal variation of 0·8 photographic magnitude in amplitude and with a period of about one day. It will be recalled that the spectroscopic period is not more than 0·71 day.

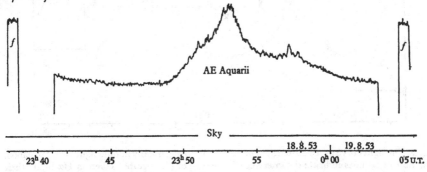

Fig. 1. Photo-electric recording of an explosion of AE Aquarii, observed through a blue filter. The deflexions at beginning and end marked *f* are of a comparison star.

An explosion is shown in Fig. 1. Three general characteristics can be distinguished in the eighty explosions observed thus far. They are as follows.

(1) The amplitude of the explosions is not constant. The strongest explosion had a photographic amplitude of 1·2 magnitude.

(2) The total duration of an explosion is about 15 minutes, this length varying little with the amplitude of the explosions. The phenomenon is thus very transient.

(3) The tracings of all the explosions show a sharp peak at maximum despite the rapid motion of the recording paper. The increasing and

* Present address: Observatoire du Pic-du-Midi, Bagnères-de-Bigorre (Hautes-Pyrénées), France.

3-2

decreasing phases of the strong explosions always show small bursts which stand out from the natural fluctuations of the anode current of the cell.

The time constant of the combined cell and recorder plays an important part in the faithful reproduction of the light curve and in the recording of the small bursts mentioned above.

In Fig. 2 is illustrated a rough regularity in the occurrence of the explosions. These generally appear when AE Aquarii has a high mean brightness. However, it is not possible to predict either the time of occurrence or the amplitude of the explosions, which often overlap one another.

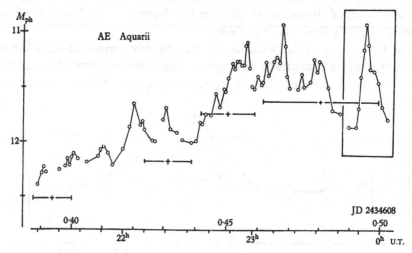

Fig. 2. The light curve of AE Aquarii, as observed photo-electrically in blue light on JD 2434608. The horizontal lines indicate the intervals of exposure of photographs taken in Hα light; their positions with respect to the ordinate axis represent the magnitude in Hα light. The rectangle encloses the explosion that is shown in the form of the original record in Fig. 1.

A deeper understanding of the explosion phenomenon requires successive observations in several colours. Such a study is difficult, and the chance of obtaining successful results is not large in the case of such an erratic pheno-menon. All of our measures show, however, that the amplitude of an explosion increases from the yellow to the ultra-violet.

Dr Joy has established the presence of strong bright lines of hydrogen. Their influence may be estimated by means of the two following photo-metric methods.

(1) Two telescopes of the Observatory of Haute Provence follow AE Aquarii simultaneously. At the 120 cm. telescope, the photo-electric photometer records AE Aquarii continuously. The second (80 cm.) telescope photographs AE Aquarii and the neighbouring stars with a

36

pseudo-monochromatic combination which is centred on Hα (by the use of 103 *a*-E plates and an Ilford 6o8 filter). The photographic exposures are made on the basis of the appearance of the photo-electric record. We have succeeded in centering the exposure times of different plates sometimes upon calm phases, and sometimes upon the explosions. Examination of the plates indicates a strengthening of the Hα emission line at each explosion. This result may be criticized because this photographic observation covers too large a spectral interval, and the length of the exposure blurs the phenomenon.

(2) Mr Ring has obtained for us an interference filter which transmits a band 70 angstroms in width. At each explosion, measures are made alternately upon a bright line and upon the neighbouring continuous spectrum. The result is the same as before: the bright lines of hydrogen contribute two-thirds of the amplitude of the explosion, the variation of the continuous spectrum contributing the remaining one-third.

AE Aquarii is a faint star, and the quality of the results is limited by the small number of photons received per second in a limited spectral region. An increase in the sensitivity would be detrimental to the rapidity of response. But an explosion is a rapid phenomenon that is difficult to record correctly as well as alternatively in two limited spectral intervals. The diversity of the amplitudes of successive explosions makes it impossible to compare the observation of an explosion, observed in an emission line with that of a second explosion observed in the adjacent spectral region. Therefore, the erratic character of the rapid explosions is the principal experimental difficulty preventing a more detailed photometric study.

II. INSTABILITY IN THE HOT STARS OF LOW LUMINOSITY AND IN THE NOVAE

7. EVIDENCE FOR INSTABILITY AMONG SUB-LUMINOUS STARS

JESSE L. GREENSTEIN

Mount Wilson and Palomar Observatories, Pasadena, California, U.S.A.

A programme of spectroscopic study of sub-luminous stars has been in progress at Palomar, using the coudé spectrograph. Various types of objects have been studied and some preliminary results are available. Certain other programmes will, when completed, be relevant to the subject-matter of this Symposium. A survey of possible velocity variations in sub-dwarfs of spectral types A to G is in progress. So far I can report that no emission lines have been found, and that there are very few spectroscopic binaries in this group. In another programme, the depth of the late-type absorption lines in the composite spectra of SS Cygni stars will be used to estimate the luminosities of the hot components near their minima. Spectra of old novae are also being obtained with a dispersion of 38 Å/mm. Plates obtained thus far show that complex structure still exists in the emission bands.

I wish to report some results on the spectra of white dwarfs and hot sub-dwarfs. As far as I know, light variations have not been found in the white dwarfs. In view of the results of photo-electric studies (H. L. Johnson and D. L. Harris) of the colours and magnitudes, it is probable that any light variation is small. The spectra of white dwarfs have now been found to show considerable variety, and the evidence for or against instability takes different forms dependent on the type of spectrum.

(1) Twelve white dwarfs with strong hydrogen lines have been studied spectrophotometrically. There is a wide range of line strength, colour and probably temperature; equivalent widths of Hγ range from 2 to 40 angstroms. In *no* case has peculiar spectral features been noted, nor has variability been suspected. Radial velocities are too poor to be useful. For the sharp-lined star L 532–81, my velocity $+57 \pm 8$ km./sec. agrees perfectly with that determined at the McDonald Observatory. But for a broad-lined star, W 1346, I find $+38 \pm 7$ km./sec.; from low dispersion the Mount Wilson result was $+101$ km./sec. and the McDonald $+26$ km./sec. Popper in his detailed study of 40 Eri B found no significant velocity variations.

(2) One strong-lined white dwarf HZ 9 (= L 1239–16) is a member of the Hyades cluster; its absolute magnitude is near + 12. There are emission lines in its spectrum: Hγ, Hδ, H + Hϵ, K, λ 3905 of Si I and Hζ. The hydrogen emissions are seen at the bottoms of the strong absorption lines, but K and λ 3905 are superposed on the continuous spectrum and must be intrinsically quite strong. The emission lines are slightly asymmetric with an absorption feature at velocity + 40 km./sec. The centre of the emission, which is 250 km./sec. wide, lies near + 70 km./sec. The velocity of the cluster is about + 36 km./sec., so that this displacement could be interpreted as an Einstein shift in the emission, which disappears in the outer absorbing material. However, asymmetric self-reversal is so common in Be stars that the velocities should not be over-interpreted. While this discovery of emission is very unexpected, an alternative explanation is possible. The colour indices of the star were measured photo-electrically by D. L. Harris who kindly communicated his results: B − V = + 0·33, U − B = − 0·69. Such a B − V colour makes it the reddest of his normal A-type white dwarfs, while the U − B corresponds to a rather hot white dwarf. He interprets the star as a composite of a dM5 star and a hot white dwarf. Luyten had previously suspected the star to be bright on an infra-red photograph. In the blue the white dwarf would be 2·2 magnitudes brighter, in the ultra-violet 4·1 magnitudes brighter than the dM star. While this analysis explains the colour indices, the strength of the emission lines is surprisingly high, if they are ascribed to the red star. For example, λ 3905 would be thirty times the strength of the dM continuum.

(3) There are few late-type white dwarfs known, but in two out of the three wF stars so far examined, an unexpected shell-like phenomenon was found in the H and K resonance lines of Ca II. Relatively sharp cores are superposed on strong, pressure-broadened lines. The star L 745–46 gave core velocities of + 24 ± 25 and − 18 ± 15 km./sec. from two plates. No other lines were measurable. The star R 640 has stronger lines, and the exposure was insufficient to reach cores, if they are present. The most detailed investigation has been made of Van Maanen 2 (= W 28). This object has very greatly broadened Fe I lines in the ultra-violet, and broad H and K with superposed sharp cores. It is probable that the 'shell' or core-producing region is variable in velocity and extent. Five old low-dispersion Mount Wilson plates had given a catalogue velocity of + 263 km./sec. No recent plate approaches this value, the range of measured velocities being from about + 30 to + 140 km./sec. Table 1 includes a journal of the recent measures.

42

Only one day elapsed between the dates of N 203 and N 205, and between those of Pe 1821 and 1824. It is probable but not certain that the velocity is variable; a sharp core might be superposed with a relative negative displacement on a broad line from the reversing layer. An increase of plate density, or more probably of the intensity of the core, could explain a reduced velocity. Setting on the centre of the broad line is impossible at high dispersion, in the presence of a deep core. The older low-dispersion work might give an average velocity. However, even with this as a tentative possibility, the apparent structure of the lines and the present velocity cannot be understood without an extended stellar envelope possessing a velocity gradient. The envelope or shell need not be detached. With a velocity of escape of 4000 km./sec., the forces producing ejection or distention must be very powerful. I can mention magnetic fields, rotation or internal instability as possibilities. Incidentally, the lower positive velocity now found makes the galactic orbit of this star less peculiar than that resulting from the catalogue value, and obviates the need for an improbably large red shift.

Table 1. *Velocity Measures in Van Maanen 2*

Plate	Dispersion (Å/mm.)	Velocity (km./sec.)	Probable error (km./sec.)	Remarks
N 203	192	+ 21	±60	Dense, core only
N 205	192	+239	40	Thin, whole width
Pd 548	18	+122	30	Very thin, whole width
Pd 1203	18	+ 48	4	Whole width
Pe 1258	38	+ 31	15	Core only
Pe 1301	38	+ 28	18	Core
		+137	18	Whole width
		+300	—	Red core? H only
Pe 1821	38	+ 54	5	Thin; cores complex?
Pe 1824	38	+ 71	±30	Thin, whole width

(4) Hot sub-dwarfs and white dwarfs with strong helium lines exist in considerable number. The very blue star $+28°$ 4211, closely related to the white dwarfs, shows strong He II and H, and no peculiar spectral features. Two white dwarfs with very strong, broad He I lines, found by Luyten, have extraordinary spectral characteristics which indicate ejection or an extended envelope. They are very blue, about apparent magnitude 15, and have $M \approx +11$. No hydrogen lines are seen. The He I lines are extraordinarily strong, and some of the $2^3P° - n^3D$ lines show relatively sharp centres. But the absolutely metastable lines, λ 3888, $2^3S - 3^3P°$ and λ 3964, $2^1S - 4^1P°$ have very much sharper and deeper cores. In fact, the sharp λ 3888 line is the strongest feature of the spectrum to the eye, although the broad shallow lines like $\lambda\lambda$ 4471, 4026 actually have twice

43

the equivalent width. It should be remembered that $\lambda\lambda$ 3888, 3964 appear in absorption in the spectra of stars in the Orion Nebula, a few other diffuse nebulae, and in 30 Doradus. The metastability of the lines under nebular conditions of the dilution of radiation is well established theoretically; however, they would be enhanced with respect to other He lines even if the radiation is only slightly dilute. The cores are not completely sharp, and may show velocity broadening. If the optically thin envelope region had a radius 10 or 100 times that of the white dwarf, any initial random mass motion with velocity near that of escape would be substantially reduced by gravity.

The spectra so far obtained are few in number, since at 38 Å/mm. a 15th magnitude star is a difficult object even with the 200-inch Hale reflector. The star L 930–80 shows, on two plates, cores at a displacement of −15 km./sec., as compared to a stellar velocity of −2 km./sec. Weak displaced components are also suspected in five He I lines, on different plates, at a velocity of −320 km./sec. The star L 1573–31 has a velocity of +45 ± 3 km./sec., from five good spectrograms. The cores of $\lambda\lambda$ 3888, 3964 are at +34 ± 7 km./sec., and sharp components of λ 3888 are suspected on two plates at −140 and −310 km./sec. It is surprising that the strong cores are so near zero velocity. The highly displaced lines are weak and not seen on all plates, but are probably real. More observations are needed, but are difficult since the star is near magnitude 15. My opinion is that both helium-rich white dwarfs have extensive envelopes.

(5) Among the faint blue stars a considerable number of very interesting O-type sub-dwarfs have been studied, for example HZ 1, 3, 44 and a brighter object found by G. Münch, HD 127493. These stars have weak H, strong He I and He II, and many sharp lines of N II, N III, Si IV, etc. Their luminosities can only be estimated; pressure and Stark broadening is large, but interstellar lines are weakly seen in the magnitude 13 objects. Most of the sharp lines, due to N III and Si IV, arise from non-metastable, highly excited levels, so that the pressure is lower than in white dwarfs but is high enough to broaden He I lines. Probably the absolute magnitudes are between +1 and +5. All four stars have sharp cores in $\lambda\lambda$ 3888 and 3964, as well as traces of cores in the less metastable He I lines of the $2^3P°$ series.

The existence of the cores apparently demonstrates that low-pressure envelopes exist with a considerable dilution of radiation. However, there are only very small velocity shifts. HD 127493 has been studied at dispersions of 10 Å/mm. and 18 Å/mm. The largest line shift found is −6 ± 2 km./sec. for λ 3964, and the mean shift of the metastable lines is −3 km./sec. In HZ 44, observed at 18 Å/mm., the mean shift is barely significant, -5 km./sec. The fainter stars, HZ 1 and HZ 3, have a non-significant shift of -3 km./sec., measured at 38 Å/mm. The stars of this O-sub-dwarf group are at moderate to high galactic latitudes, and have a velocity dispersion, as a group, slightly larger than do the ordinary B stars. HZ 3 has the highest radial velocity ($+43$ km./sec.), but correction for normal solar motion leaves a radial component of the space motion of only $+25$ km./sec.

In conclusion, the helium-rich white dwarfs and sub-dwarfs show no evidence of a large velocity gradient from the reversing layer upwards, but all have sharp cores in most metastable lines. No astrophysical explanation for the sharp core lines exists except a very large pressure gradient in the atmosphere, and the relative enhancement of the cores of the most metastable lines would suggest that the atmosphere is large compared to the star. Observations designed to detect possible short-period light variability would be desirable.

8. THE EXTREMELY RAPID LIGHT-VARIATIONS OF OLD NOVAE AND RELATED OBJECTS

MERLE F. WALKER

Mount Wilson and Palomar Observatories, Pasadena, California, U.S.A.

It has been recognized in recent years that extremely rapid light-variations occur in certain stars. Flares lasting only a few minutes have been observed in a number of dMe stars [1]. Flaring has also been observed in the W Ursae Majoris star U Pegasi [2]. Rapid variations in light have been found in some T Tauri stars [3], in the U Geminorum variable AE Aquarii [4, 5], and in the short-period eclipsing binary UX Ursae Majoris [6, 7, 8]. Observations during the past two years at the Mount Wilson Observatory have shown that extremely rapid and apparently continuous variations in light occur also in many of the old novae and in a number of related stars.

The investigation of these stars was undertaken following the discovery of rapid light-variations in MacRae $+43°$ 1 [9], whose spectrum resembles that of an old nova [10]. The light-variations of MacRae $+43°$ 1 appear to be completely random and have cycle lengths ranging from 1 to 30 min. The star has been observed to vary in ultra-violet light by as much as 0·4 magnitude in 5 min. The range of the fluctuations varies with wavelength, the amplitude in the yellow and blue being 0·7 and 0·8, respectively, of the amplitude in the ultra-violet.

To date, thirteen old novae, three possible novae, five nova-like variables, four U Geminorum stars, two binary Mira-type variables, and one white dwarf have been examined for rapid variations in light. Some traces of short-period variability similar to that of MacRae $+43°$1 were found in all of the novae, the possible novae, the U Geminorum stars, and the Mira-type variables. Little or no variation was found in four of the nova-like variables or in the white dwarf. The details of the observations are given in Table 1. Since the present programme was of an exploratory nature, not enough observational material has been obtained for most of the stars to permit a complete analysis of the light-variations. Consequently, the activity of the stars has been described in Table 1 merely by giving for each object the maximum variation in light during the period of observation, the dispersion of points measured at equal time intervals

46

along the light-curve about the level of mean light during the observations, and the average time interval between maxima.

Table 1 indicates that the most active of the old novae is T Coronae Borealis. Sections of Brown Recorder sheets showing the light-variations of this star in the ultra-violet on 30 May and 1 June 1954 are reproduced in Figs. 1 and 2, respectively. Near ultra-violet spectra having a dispersion of 83 Å/mm. were obtained by A. J. Deutsch during several of the photo-electric runs on T Coronae. These spectrograms would have shown any pronounced changes in the line spectrum that might have accompanied the variations in light. Since no changes in the appearance of the spectrum were found, the light-variations cannot be attributed to fluctuations in the strength of emission lines or in the continuous Balmer emission. The light-variations must result from fluctuations, presumably of a thermal nature, in the strength of the continuous spectrum. During the spring of 1954, the brightness of the nova component was such that it dominated the blue and violet regions of the spectrum. It was found that even though the amplitude of light-variation was very large in the blue and ultra-violet, there was practically no trace of activity in yellow light. This, together with the spectroscopic results, is consistent with the hypothesis that (a) the system is composed of two distinct stars, and (b) only the nova component is responsible for the rapid variations in light.

The other old novae that were observed show varying amounts of activity. Two of them, Nova (T) Aurigae 1891 and Nova (WZ) Sagittae 1913, 1946, do not appear to show the extremely rapid fluctuations having periods of only a few minutes that were found in the others. These two stars, while they are definitely variable, behave in a much more leisurely fashion, exhibiting maxima only every 40 or 50 min. The degree of variability of the old novae does not appear to be correlated with the type of nova or the time elapsed since the outburst. There is also no correlation with the appearance of the spectrum, as given by M. L. Humason [11]. However, the spectra may have changed since his observations were made; to be certain whether or not such a correlation exists, spectra would have to be obtained concurrently with the photometric observations. There may be a correlation between the degree of short-period activity and the presence of long-period variations in light. The stars with the greatest amounts of short-period activity tend to be the ones which have been found to be variable from visual observations. It must be remembered that at present we know nothing about the 'secular' occurrence of these short-period variations. It may be that the differences in activity found in the different stars result primarily from long-term changes in the degree of activity of all the stars.

47

Table I. Observations of Old Novae and Related Stars

Type	Star	Date (U.T.)	V (mag.)	B–V (mag.)	U–B (mag.)	Maximum variation during observation (mag.)	Dispersion (mag.)	Average interval between maxima (min.)	Interval observed (min.)	Notes
Novae	Nova (GK) Per 1901	1953 Oct. 28	12·96	+0·69	−0·56	0·20	—	(4·2)	21	1
		29	12·96	+0·74	−0·60	0·15	0·035	5·0	52	—
	Nova (T) Aur 1891	1954 Feb. 24	14·92	+0·28	−0·64	0·14*	0·039*	(50·)	50	2
	Nova (DN) Gem 1912	1953 Oct. 29	15·56	+0·20	−0·95	0·08	—	—	—	1
		Nov. 2				0·06			25	1
		30				0·05		(5·)	—	1
	Nova (V481) Oph 1848	1954 Feb. 25	13·47	+0·39	−0·57	0·05	—	—	14	1
		May 31	13·52	+0·36	−0·64	0·03	—	—	—	3
	Nova (DQ) Her 1934†	1954 Aug. 29	14·17	0·00	−0·88	—	—	—	—	4
	Nova (V603) Aql 1918	1954 April 2	11·58	−0·09	−0·81	0·26	0·066	3·2	66	3
		May 31	11·50	−0·04	−0·82	—	—	—	—	3
		July 8		+0·08	−0·86	0·11*	0·034*	(10·)	36	—
	Nova (HR) Lyr 1919	1954 July 8	15·71	+0·15	−0·94	0·30	0·078	10·2	50	3
	Nova (V476) Cyg 1920	1954 July 9	17·09	+0·33	−0·65	0·28	0·096	11·2	40	—
	Nova (Q) Cyg 1876	1953 Oct. 31	14·93	+0·24	−0·64	0·05	—	—	48	3
	Nova (DI) Lac 1910	1954 Aug. 30								—
		1953 Oct. 30	14·57							1
Recurrent novae	T CrB†	1954 May 29	—			0·49	0·106	3·5	106	5
		June 1	—			0·28	0·072	4·0	100	—
	RS Oph	1954 May 31	11·28	+1·14	−0·04	0·13	0·034	(3·)	12	3
		July 31	10·74	+1·07	−0·06					3
		July 8								—
	Nova (WZ) Sge	1954 July 8	15·21	+0·09	−0·71	0·24*	0·065*	(40·)	41	3
Possible novae	V426 Oph	1954 May 30	12·64	+0·50	−0·74	0·30	0·082	3·4	31	—
		July 8								—
	MacRae +43° 1†	1953 Aug. 7				0·44	0·102	3·2	55	3
		9	13·38	−0·13	−1·05					—
		9	13·67	−0·08	−0·98					—
	EM Cyg	1954 May 31	12·90	+0·17	−0·72	0·24	0·048	6·3	19	—
		June 1				0·18	0·069	7·2	43	—

Nova-like	V Sge	1954 May 30	—	—	—	0·08	0·022	(11·)	22	—
		July 8	12·24	+0·03	−0·87	—	—	—	—	3
	BF Cyg	1954 July 30	—	—	—	—	—	—	19	5, 6
	CI Cyg	1954 July 30	—	—	—	—	—	—	8	5, 7
	P Cyg	1953 Nov. 12	—	—	—	—	—	—	20	8
	Z And	1954 Aug. 29	—	—	—	0·02	0·006	(13·)	26	5
U Gem	RX And	1954 July 31	—	—	—	0·15	0·032	2·1	41	—
	AE Aqr†	1954 July 30	—	—	—	1·51	0·371	5·7	103	—
	SS Cyg†	1954 July 9	11·90	+0·62	−0·62	0·22	0·057	3·3	22	—
		31	—	—	—	0·12	0·033	2·7	19	—
	RU Peg	1954 July 30	—	—	—	0·06	0·018	6·4	19	—
Mira	o Cet†	1955 Aug. 11	—	—	—	0·10	0·031	(10·)	19	5, 9
	R Aqr	1954 July 30	—	—	—	0·04	0·012	(10·)	20	5, 10
White dwarf	40 Eri B	1953 Nov. 12	—	—	—	—	—	—	16	7

* Observations made in blue light; others are in the ultra-violet.   † Additional observations are available.

1. Few observations.
2. Observations poor.
3. Magnitude and colour measurement only.
4. Nova (DQ) Her 1934. Eclipsing binary. Intrinsic variations occur throughout the 4-hour cycle of the star with periods from 1 to 40 min. and amplitudes up to 0·15 magnitude in the ultra-violet. Between phases $0·110P$ (the end of eclipse) and $0·325P$, there occur extremely rapid variations of a periodic nature in the light. The range of these variations is 0·07 magnitude in the ultra-violet, and their period is 1·180 minutes. Dispersions and average intervals between maxima have not been computed for DQ Her since, owing to the unique character of this system, the source of the variations may be entirely different from that producing the variations in the other old novae.

5. Composite spectrum.
6. Possible ultra-violet variation of 0·01 magnitude.
7. No ultra-violet variation $\geqslant 0·01$ magnitude.
8. No ultra-violet variation $\geqslant 0·005$ magnitude.
9. o Cet. This star was observed when the red component was near minimum light (phase$=0·57P$). According to A. J. Deutsch (private communication), the blue star dominated in the blue and ultra-violet regions of the spectrum at the time of this observation. The observed activity thus refers to the blue component.
10. R Aqr. Observed somewhat after minimum light (phase$=0·68P$). The small amplitude found in the ultra-violet may thus result from dilution of the ultra-violet light of the blue star by light from the red component.

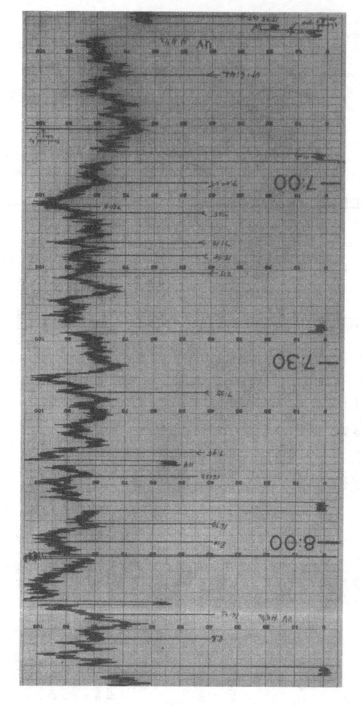

Fig. 1. The light-curve of T Coronae Borealis on 30 May 1954 (U.T.). A section of the Brown Recorder sheet showing the continuous variations of the star in ultra-violet light. Universal Time is marked at intervals of 30 min.

50

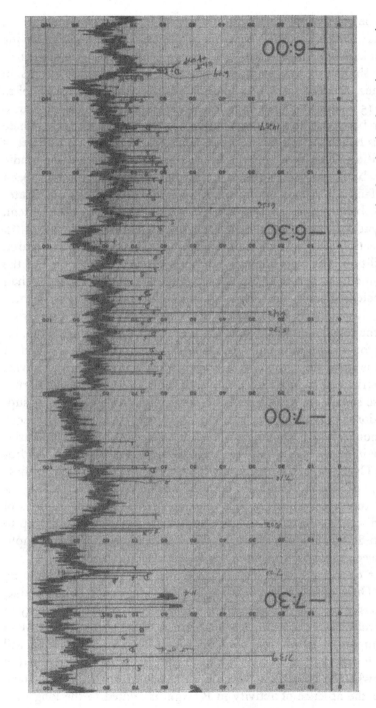

Fig. 2. The light-curve of T Coronae Borealis on 1 June 1954 (U.T.). A section of the Brown Recorder sheet showing the continuous variations of the star in ultra-violet light. Universal time is marked at intervals of 30 minutes. The level of the sky intensity is indicated by the heavy line at the bottom of the figure. In this and the preceding figure, the vertical lines depending from the light-curve are timing marks, and were not produced by the star.

51

4-2

One of the most important results of these observations of old novae was the discovery that Nova (DQ) Herculis 1934 is an eclipsing binary having the shortest known period, $4^h 39^m$. In addition to the eclipse, rapid intrinsic variability similar to that found in the other old novae also occurs in DQ Herculis. Periods from 1 to 40 minutes and ultra-violet ranges of as much as 0·15 magnitude have been observed. In addition, periodic varia- tions occur between phases 0·110P and 0·325P in the eclipse cycle. These fluctuations have a period of 1·180 min. and a range in the ultra-violet of 0·07 magnitude. This is the only star in which rapid variations of a periodic nature have been observed. While no explanation of these variations has been found, the fact that they occur only during a certain phase interval indicates that they probably result in some manner from the binary nature of the system. The observations of this system have been discussed in detail elsewhere [12].

The detection of rapid light-variations in the old novae led to a search for variability of this kind among a number of related stars. One of the first experiments undertaken was the investigation of two stars that had been suspected of being old novae from the appearance of their spectra. EM Cygni was found by E. M. and G. R. Burbidge [13] to have a spectrum similar to that of an old nova, while a similar result was obtained by G. H. Herbig [14] for V426 Ophiuchi. Rapid light-variations were detected in both of these stars, strengthening the supposition that they are old novae.

Rapid variations in light have also been found in four U Geminorum stars. These stars were included in the programme because of their sup- posed relationship to the novae and since the observations of K. Henize [4] and F. Lenouvel and J. Daguillon [5] had already shown that rapid varia- tions occur in one of them, AE Aquarii. In the other three, RX Andro- medae, SS Cygni, and RU Pegasi, the light-variations are similar to those found in the old novae. However, both the work of Lenouvel and Daguil- lon [5] and the present observations indicate that the light variations of AE Aquarii consist primarily of large outbursts having amplitudes up to 1·5 magnitude, upon which are superimposed fluctuations of lesser ampli- tude. Between these outbursts the star is relatively quiescent.

Four out of the five nova-like variables observed showed little or no variation. Three of these four, BF Cygni, CI Cygni, and Z Andromedae, have composite spectra. Consequently, the failure to detect variability in these stars may have resulted from the faintness of the blue star relative to the red component at the time of observation. Such an explanation will not suffice for the fourth star, P Cygni. Further work will be required to determine whether rapid light-variations are absent in all P Cygni stars, or whether the absence of activity in P Cygni is related to the long time interval which has elapsed since the outburst of this star. The elapsed time since the explosion of P Cygni is much longer than for any of the old novae examined in this programme.*

The observations of the two Mira-type variables were made near the light-minima of their red components. In o Ceti, the spectrum of the blue star was dominant in the blue and ultra-violet at the time of the photo-electric measurements, according to spectroscopic observations by A. J. Deutsch. R Aquarii was observed somewhat further from minimum light than was o Ceti, and it is therefore possible that the small amplitude found in this star may have resulted from the effect of ultra-violet light from the red component.

While no detectable light-variation was found in the white dwarf 40 Eridani B, this result is not completely conclusive since the star was observed for only a short time on a single night.

Magnitudes and colours for most of the stars have been obtained on the U, B, V system of Johnson and Morgan [15]. These result mainly from single observations or means of only a few observations obtained while the stars were being monitored for rapid variations in light, and the results may differ very appreciably from the true mean values. For some of the more active stars, even these 'instantaneous' colour measurements may be in error by several hundredths of a magnitude owing to changes in the brightness of the star during the colour observations.

The U–B and B–V colours of the old novae are plotted in Fig. 3. It will be seen that these stars deviate markedly from the relationship between the two colour indices found for normal main sequence stars, which is indicated by the curved line. Such a departure could result from interstellar reddening, abnormal brightness of the star in the ultra-violet, or a composite spectrum. The last possibility may be ruled out on the basis of the spectroscopic observations of these stars. One possible exception might be RS Ophiuchi, which was reported to have had in 1923 an absorption spectrum corresponding to a spectral class of about G 5 [16]. The two diagonal lines in Fig. 3 represent reddening paths, and the fact that the old novae fall roughly along a sequence parallel to these lines might suggest that their abnormal colours result from interstellar reddening. The amount of reddening of a few of the old novae can be checked roughly since they occur in fields where the absorption has been derived from star counts. Thus, Nova (GK) Persei 1901 is reddened by at least 0·2 magnitude [17]. The amount of reddening is probably larger than this since observations of the 'light echo' seen after the 1901 maximum indicate that the star must be

---

* Note by the Editor: The detection of small variations in the light of P Cygni was reported by Kharadze at the Symposium: see p. 87.

situated in or near a dark cloud. On the other hand, Nova (T) Aurigae 1891 is located in a region of very light obscuration[18]. It would appear likely that the location of the stars in the two-colour diagram results from a combination of reddening and abnormal ultra-violet intensity. Additional work on this problem is needed.

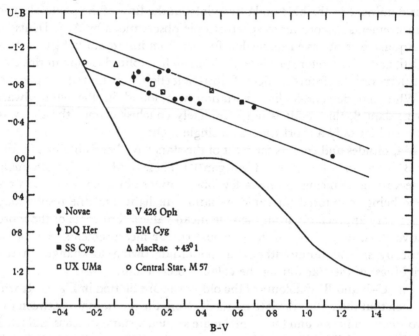

Fig. 3. Two-colour diagram of old novae and related stars. The relationship between the colours for un-reddened main sequence stars is shown by the heavy curved line. The two diagonal lines represent reddening paths.

The colours of the central star of the planetary nebula M 57 have been plotted in Fig. 3 for comparison with the colours of the old novae. Unlike the novae, the colours of this star appear to be normal. If the observed colours of the novae represent an intrinsic abnormality in the energy distributions of these objects, then the differences in the colours may indicate that the novae are quite different from the nuclei of planetary nebulae. However, if the colours of the old novae are intrinsically normal and their position in Fig. 3 is the result of interstellar reddening, then the colours and temperatures of the old novae must be very similar to those of the central star of M 57.

It is to be noted that the colours of the three stars suspected of being novae from their spectra, V 426 Ophiuchi, MacRae +43° 1, and EM Cygni, agree quite well with those of the known novae.

Much additional work will be needed to determine the cause and nature of the rapid light-variations. However, certain conclusions may be drawn from the present material. As we have pointed out, the dependence of the amplitude of the light-variations upon wave-length and the fact that in T Coronae there are no changes in the line spectrum accompanying the variations in light indicate that the fluctuations are produced by variations in the temperature of some portion of the surface of the star. The observations also suggest that the disturbances producing the variations involve a large portion of the surface of the star. For example, the temperature of MacRae +43° 1 corresponding to the B–V colour is about 13,000° K. The largest observed short-period variation in ultra-violet light was 0·44 magnitude. If this light-variation were produced by a spot or 'flare' on the surface of the star, then if the temperature of the spot were 100,000°, the radius of the spot would have to be 0·11 that of the star, while if the temperature of the spot were 1,000,000°, the radius of the spot would still have to be 0·03 the radius of the star. If the intrinsic colours of MacRae +43° 1 and of the old novae are normal and their observed colours the result of interstellar reddening, then the situation becomes even worse. If the temperature of the star is actually 25,000°, then if the temperature of the spot is 100,000°, its radius must be 0·27 that of the star, while if its temperature is 1,000,000°, the radius will have to be 0·07 of the radius of the star. On the other hand, if the observed colours of the old novae are not affected by reddening, the spot can be somewhat smaller. If the temperature of the star is actually 8,000° and the temperature of the spot is 100,000°, its radius would be 0·026 that of the star, while if the spot temperature were 10,000°, its radius would be 0·082 that of the star.

Perhaps we might suppose as a working hypothesis that the rapid light-variations are caused by the alternate heating of the atmosphere of the star by shock waves and its subsequent cooling by radiation. If this is the case, then the rapidity of the light-variations would suggest that the quantity of material which is being heated and cooled is relatively small, and consequently, if the entire surface of the star is involved, that the star itself is small.

What is the origin of the observed instability of these stars? The fact that different classes of stars display this same type of rapid light-variation suggests that there may be some sort of relationship between them. It is interesting to note that a large number of these stars are known to be binaries: Nova (DQ) Herculis has been shown to be an eclipsing binary. AE Aquarii has been shown by A. H. Joy[19, 20] to be a spectroscopic binary, and Joy[19] has also found that SS Cygni and RU Pegasi have variable radial velocities and may thus be binaries. o Ceti is a visual double, while the duplicity of T Coronae, Z Andromedae, and R Aquarii is indicated by the composite spectra of these objects. In addition, rapid light-variations occur in the short-period eclipsing binary UX Ursae Majoris. Thus, it is possible that all of the novae and related stars are binaries and that both the short-period fluctuations and the more violent explosions of these objects result in some way from their binary nature. Additional observations will be required to determine whether or not this is actually the case. Even if all of these stars are binaries, it is probable that binary nature *per se* is not both a necessary and sufficient condition for the occurrence of novae or nova-like phenomena. This is shown by the fact that all of the novae fall in a restricted area of the Hertzsprung-Russell diagram.

In conclusion, it should be noted that recent theoretical investigations[21, 22] indicate that white dwarf stars will be unstable if sources of nuclear energy are present in them. Consequently, we might expect that white dwarfs would show no fluctuations in light, in accord with the observations of 40 Eridani B. On the other hand, if it is true that the novae represent stars which are approaching the white dwarf stage, it might be that the rapid variations, and perhaps the more violent outbursts, indicate that the star is in a stage where it is becoming unstable with respect to the processes of nuclear energy generation, but in which these processes have not yet become entirely exhausted.

REFERENCES

[1]   A list of the known flare stars is given by P. E. Roques, *Publ. A.S.P.* **67**, 34 (1955).
[2]   M. Huruhata, *Publ. A.S.P.* **64**, 200 (1952).
[3]   G. Haro and W. W. Morgan, *Ap.J.* **118**, 16 (1953).
[4]   K. G. Henize, *A.J.* **54**, 89 (1949).
[5]   F. Lenouvel and J. Daguillon, *Ann. d'Ap.* **17**, 416 (1954).
[6]   A. P. Linnell, *Harvard Circ.* No. 455 (1950).
[7]   H. L. Johnson, B. Perkins and W. A. Hiltner, *Ap.J. Suppl.* **1** (No. 4), 91 (1954).
[8]   M. F. Walker and G. H. Herbig, *Ap.J.* **120**, 278 (1954).
[9]   M. F. Walker, *Publ. A.S.P.* **66**, 71 (1954).
[10] J. L. Greenstein, *Publ. A.S.P.* **66**, 79 (1954).
[11] M. L. Humason, *Ap.J.* **88**, 228 (1938).
[12] M. F. Walker, *Ap.J.* **123**, 68, (1956).
[13] E. M. and G. R. Burbidge, *Ap.J.* **118**, 349 (1953).
[14] G. H. Herbig, private communication.
[15] H. L. Johnson and W. W. Morgan, *Ap.J.* **117**, 313 (1953).
[16] W. S. Adams, M. L. Humason and A. H. Joy, *Publ. A.S.P.* **39**, 365 (1927).
[17] D. S. Heeschen, *Ap.J.* **114**, 132 (1951).
[18] S. W. McCuskey, *Ap.J.* **88**, 209 (1938).
[19] A. H. Joy, *Publ. A.S.P.* **55**, 283 (1943).
[20] A. H. Joy, *Ap.J.* **120**, 377 (1954).
[21] T. D. Lee, *Ap.J.* **111**, 625 (1950).
[22] L. Mestel, *M.N.* **112**, 583, 598 (1952).

9. PHYSICAL PROCESSES IN NOVAE

E. R. MUSTEL

Crimean Astrophysical Observatory, Pochtovoje, Crimea, U.S.S.R.

The present communication is devoted to some important problems connected with the interpretation of spectroscopic phenomena observed during the outbursts of novae.

It is known that the phenomenon of an outburst in its initial stages is connected with the expansion of the 'photosphere' and the 'reversing layer' of a nova. There are some reasons to expect that this process does not involve the most central parts of the star. The increase of the radius of the 'photosphere' of the star can be estimated from the equation:

$$M_v = \frac{29500}{T} - 5 \log R_p - 0.08, \qquad (1)$$

while the expansion of the 'reversing layer' can be studied by means of integrating the curve of radial velocities determined from the displacement of lines of the pre-maximum spectrum of the star:

$$R = R_0 + \int_{t_0}^{t} V \, dt. \qquad (2)$$

From equations (1) and (2), A. Beer[1] concluded that the radius of the 'reversing layer' of Nova Her 1934 during its expansion had exceeded many times the radius of its photosphere. The author showed[2] that the same was true for each of seven novae which were observed previous to their light maximum.* The comparison of the velocities of expansion of the 'photosphere' V_p, estimated from equation (1), with the velocities V made it possible for the author[2, 3] to conclude that previous to the moment t_m of light maximum, the velocity gradient in the expanding envelope of the nova (at least up to the level where the optical depth $\tau \approx 1$) cannot be large. This is confirmed by the relative narrowness of the absorption lines in the pre-maximum spectrum of the star, previous to its light maximum.

The moment of light maximum t_m is an extremely important transitional stage in the process of the evolution of a nova. Appreciable changes in the

---

* The problem of applying formula (1) to novae has been considered by the author[2]. It was shown that this formula cannot give large errors.

spectrum of the star begin immediately after light maximum [4]. (a) The pre-maximum spectrum of the star is replaced by the principal spectrum with a greater displacement. From its very beginning, the system of the principal spectrum appears to be detached from the system of the pre-maximum one (see Fig. 1). The intensity of the systems is represented in this Figure by the breadth of the corresponding strip. For Nova Per 1901,

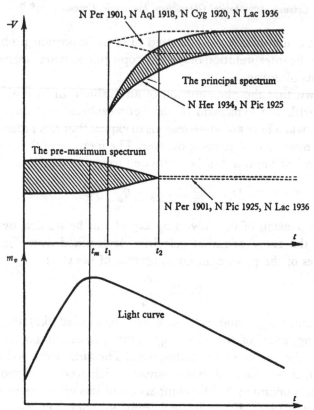

Fig. 1. The development of the principal absorption spectrum in a number of novae in the neighbourhood of maximum light.

Nova Aql 1918, Nova Cyg 1920, Nova Lac 1936 the variation of the displacement of the principal spectrum in the interval t_2-t_1 has been studied insufficiently. (b) Emission bands of the principal spectrum appear. (c) New systems known as the diffuse-enhanced spectrum and the Orion spectrum appear. Both these systems are accompanied by emission.

To explain the existence of light maximum, with which all these phenomena are intimately connected, obviously only two hypotheses A and B can be suggested [5, 6]. Two possibilities are foreseen by hypothesis A.

Hypothesis A, first model. At the moment of its outburst the star throws off a spherical shell, which begins to expand in the form of an envelope, detached from the star (Fig. 2). At first the optical thickness of the envelope satisfies the inequality $\tau \gg 1$. As long as this inequality is realized, the brightness of the star increases. While moving away from the star the envelope becomes rarefied and its τ decreases. When $\tau \approx 1$ the moment of light maximum sets in, after which the brightness of the star begins to decrease.

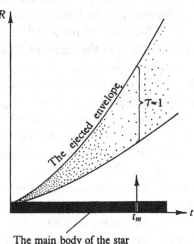

Fig. 2. A representation of the ejection of a shell by the nova under hypothesis A, first model.

This hypothesis meets considerable difficulties, as follows:

(*a*) It fails to explain the replacement of the pre-maximum spectrum by the principal spectrum, as well as all the peculiar properties which characterize this process.

(*b*) According to hypothesis A, immediately after light maximum $\tau < 1$ in the visual and the photographic regions of the spectrum. At the same time beyond the limit of the Balmer series, where the absorption coefficient is considerably greater, this inequality must be realized later. Thus, immediately after light maximum, a jump in intensity is expected to arise at the limit of the Balmer series. In other words $I(\lambda \leqslant 3647)$ would be considerably greater than $I(\lambda \geqslant 3647)$, which contradicts the observations [2]. (The normal intensity of the Balmer lines in the spectra of novae previous to light maximum suggests a quite normal hydrogen content in the second quantum state.)

(*c*) Let t_e be the moment of appearance of the emission bands in the spectrum of the star. At this moment, the inequality $\tau \ll 1$ must be realized

59

in the optical range of the spectrum. Otherwise the red wings of the bright bands, which correspond to the receding hemisphere of the envelope, would be appreciably weakened, which contradicts the observations. Assuming now that hypothesis A is valid, then at the moment t_e and even somewhat earlier we would be able to see through the detached envelope the outer layers of the central star. On the other hand, it seems obvious that whatever is the mechanism of ejection, the temperature of these outer layers must be sensibly higher than that of the detached envelope (particularly of its outer layers). Otherwise it is difficult to explain the origin of the impulse gained by the outer layers of the nova at the moment of its outburst.

Thus it should be expected that at the moment t_e the observed temperature of the nova is essentially higher as compared with its temperature at the moment t_m, which is contradicted by the observations. The latter show that in the course of a short period of time $\Delta t = t_e - t_m$, the spectrum for all novae become later. Moreover, immediately after light maximum bands of CN were observed in the spectra of Nova Aql 1918 and Nova Her 1934. Emission bands connected with the principal spectrum which appear at the moment t_e correspond exactly to the same (relatively low) state of excitation and ionization that characterizes the principal absorption spectrum. Finally, in the course of the same period of time Δt, the colour temperature of the star usually decreases also. Only a few days after the time t_e we observe a slow decrease of temperature accompanied by a transition of the spectrum of the star towards earlier subdivisions.

Hypothesis A, second model. Here the 'main burst' is followed by continuous expulsion of matter at a steadily decreasing rate. The matter ejected as a result of such a continuous process expands and its transparency increases. This model is shown in Fig. 3. The thick continuous curve represents the change of R_p for Nova Aql 1918 calculated [5] by means of equation (1) and roughly corresponds to the level with $\tau \approx 1$. Beyond this level outwards the matter is transparent. The temperatures used here are based on the spectral types of the star during its expansion. The density of shading corresponds to the density of matter. Before the outburst the value of R_p for Nova Aql 1918 did not exceed $0.4 R_\odot$. This model retains the first difficulty of the first model, whereas the second and third difficulties no longer exist. But at the same time a new difficulty arises. From elementary calculations and from Fig. 3 it may be concluded that for times before light maximum the velocity of matter (thin lines A–A in Fig. 3) must be much greater than the velocity V_p. This contradicts available data [3].*

---

* Especially if we take into consideration the influence of interstellar absorption upon M_v of formula (1); this was not done in reference 3.

in the optical range of the spectrum. Otherwise the red wings of the bright bands, which correspond to the receding hemisphere of the envelope, would be appreciably weakened, which contradicts the observations. Assuming now that hypothesis A is valid, then at the moment t_e and even somewhat earlier we would be able to see through the detached envelope the outer layers of the central star. On the other hand, it seems obvious that whatever is the mechanism of ejection, the temperature of these outer layers must be sensibly higher than that of the detached envelope (particularly of its outer layers). Otherwise it is difficult to explain the origin of the impulse gained by the outer layers of the nova at the moment of its outburst.

Thus it should be expected that at the moment t_e the observed temperature of the nova is essentially higher as compared with its temperature at the moment t_m, which is contradicted by the observations. The latter show that in the course of a short period of time $\Delta t = t_e - t_m$, the spectrum for all novae become later. Moreover, immediately after light maximum bands of CN were observed in the spectra of Nova Aql 1918 and Nova Her 1934. Emission bands connected with the principal spectrum which appear at the moment t_e correspond exactly to the same (relatively low) state of excitation and ionization that characterizes the principal absorption spectrum. Finally, in the course of the same period of time Δt, the colour temperature of the star usually decreases also. Only a few days after the time t_e we observe a slow decrease of temperature accompanied by a transition of the spectrum of the star towards earlier subdivisions.

Hypothesis A, second model. Here the 'main burst' is followed by continuous expulsion of matter at a steadily decreasing rate. The matter ejected as a result of such a continuous process expands and its transparency increases. This model is shown in Fig. 3. The thick continuous curve represents the change of R_p for Nova Aql 1918 calculated [5] by means of equation (1) and roughly corresponds to the level with $\tau \approx 1$. Beyond this level outwards the matter is transparent. The temperatures used here are based on the spectral types of the star during its expansion. The density of shading corresponds to the density of matter. Before the outburst the value of R_p for Nova Aql 1918 did not exceed $0 \cdot 4 R_\odot$. This model retains the first difficulty of the first model, whereas the second and third difficulties no longer exist. But at the same time a new difficulty arises. From elementary calculations and from Fig. 3 it may be concluded that for times before light maximum the velocity of matter (thin lines *A–A* in Fig. 3) must be much greater than the velocity V_p. This contradicts available data [3].*

---

* Especially if we take into consideration the influence of interstellar absorption upon M_v of formula (1); this was not done in reference 3.

the displacement of the principal spectrum, but not the displacement of the pre-maximum one, which corresponds to the expansion velocity of the nebulosity observed later (note particularly the cases of Nova Per 1901, Nova Her 1934 and Nova Pic 1925). As long as this nebulosity contains the main mass of matter ejected by the nova as a result of the outburst, an important conclusion can be drawn: that the envelopes of novae are

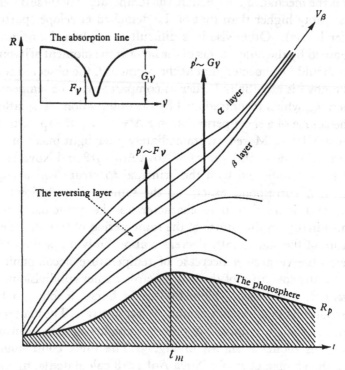

Fig. 4. A representation of hypothesis B, in which the outer parts of the extended reversing layer of the nova, as the star contracts, continue to rise in the form of an envelope.

detached after light maximum, when the principal spectrum is appearing. Moreover, because we can see the birth of the principal envelope itself (the appearance of an extremely weak principal spectrum at the very beginning) it means that the latter originates inside the extended reversing layer of the nova, which is transparent for the frequencies of the continuous spectrum.

The difficulties of hypothesis A force us to consider hypothesis B.

Hypothesis B. It is supposed that the 'photosphere', which determines the brightness of the nova, has its greatest dimensions at the moment of light maximum, after which the contraction of the photosphere begins, followed by a decrease of brightness of the nova. From the point of view of this hypothesis the matter which forms the 'photosphere' remains with the star.* The outer parts of the extended reversing layer of the nova leave the star in the form of an envelope. As was shown by the author in references 6, 8, 9, and 10, hypothesis B permits one to explain the process of replacement of the pre-maximum spectrum by the principal spectrum.

This process can be easily understood from Fig. 4. The rise of a large velocity gradient in the inner parts of the extended reversing layer of the star near the moment t_m makes these parts transparent even for the frequencies of absorption lines. This leads to an almost sudden increase of selective radiation pressure upon the internal parts of the envelope, which is in a state of detachment and has a small gradient of velocities. As a result of this a new β layer is formed which moves away from the star. This β layer creates absorption lines of the principal spectrum. The 'old' envelope with a small gradient of velocities decreases in thickness. This envelope, called the α layer, corresponds to the pre-maximum spectrum. When the β layer reaches the outer parts of the α layer, the latter disappears and the pre-maximum spectrum disappears with it. There remains only the completely detached principal envelope, which can be observed later in the form of a bright nebulosity. This mechanism offers a possibility of estimating the thickness ΔR of the envelope which is being detached. It is obvious that

$$\Delta R = \int_{t_1}^{t_2} |V_\beta - V_\alpha| \, dt. \tag{3}$$

The moments t_1 and t_2 are fixed in Fig. 1. Computations show [6] that for Nova Her 1934 we have $\Delta R \approx 60 \, R_\odot$ for metallic lines, and $\Delta R \approx 100 \, R_\odot$ for hydrogen lines.

The above mechanism explains furthermore all the peculiar properties which characterize the replacement of the pre-maximum spectrum by the principal spectrum, namely:

(a) The formation (after light maximum) of subsidiary absorption systems with displacements that are larger than the displacement of the principal spectrum.

(b) The increase of displacement of the principal spectrum at the interval $t_2 - t_1$ (see Fig. 1).

(c) The existence of a very weak pre-maximum spectrum after light maximum. This weak spectrum was observed for Nova Per 1901, Nova Pic 1925 and Nova Lac 1936. A special investigation carried out by the

---

* A few possible cases of the motion of matter in hypothesis B are considered in reference 6, pp. 116–20.

author[9] has shown that all the other mechanisms which were suggested to explain the replacement of the pre-maximum spectrum by the principal spectrum involve very serious difficulties.

It is not possible to agree with the suggestion of McLaughlin[4] that the replacement of the pre-maximum spectrum by the principal spectrum is connected with the increase of the radiation pressure caused by a sudden increase of temperature of the star. We have seen above that there is no increase of temperature immediately after light maximum. Moreover, for some stars (for example, Nova Her 1934) the replacement of the pre-maximum spectrum by the principal spectrum took place even before the appearance of the emission bands.

From all that has been said above, it follows that from the spectroscopic point of view, hypothesis A meets a number of extremely serious difficulties which do not exist in B. Hypothesis B permits one also to explain the replacement of the pre-maximum spectrum by the principal spectrum. Such a replacement was observed in all the seven novae 'caught' before maximum and it ought to be considered as a fundamental law in the spectroscopic evolution of every nova[4].

Let us now examine the consequences which follow from hypothesis B. If this hypothesis is valid, then in this case the photospheric matter must remain with the star. Assuming that the nature of forces which retain the photosphere is gravitational, we can estimate then the minimum masses of novae. In fact, we may consider that the velocity V_p of the photosphere, which can be derived from equation (1), is less than the parabolic one for the level R_p. Extremely large masses are obtained in this case[5], which are greater, the greater the luminosity of the nova previous to its outburst. Of all the obtained masses the largest is the mass of Nova Aql 1918, for which $m \approx 1700\,m_\odot$, and the smallest mass is the one of Nova Her 1934, for which $m = 7\,m_\odot$.

The cases in which displacement of the pre-maximum spectrum diminished with time also gave large 'masses' of novae. It is true that a number of authors have raised the question that probably there was no real deceleration of matter here, but simply a certain movement of some 'effective' absorbing level in the reversing layer of a nova. However, a critical investigation of this problem made the author arrive at the conclusion[11] that in certain cases there had been a real deceleration of moving matter. Then again the 'masses' reach several hundred solar masses.

In this connexion the following facts may also be mentioned. R. F. Sanford[12] discovered that the absorption lines of the type-gM3 component of the recurrent Nova T CrB show orbital motion with a period of about 230 days and a semi-amplitude of 21 km./sec., while the emission bands in the spectrum of the nova itself do not show any periodic variations. This case, for the present unique in the history of stellar spectroscopy, may be explained by the fact that the mass of T CrB exceeds considerably the mass of its giant satellite. Further, it seems of interest to note that the 'masses' of novae, estimated by means of the above-mentioned methods, are proportional to the masses of the envelopes ejected by the novae. The values of the masses of the envelopes were determined by I. Kopylov [13] for a number of stars by means of the same method.

The introduction of large masses for novae is connected, however, with extremely serious difficulties:

(a) The emission bands observed in the spectra of post-novae do not show the presence of large gravitational displacement expected in this case.

(b) According to recent data, Nova Her 1934 is an eclipsing variable [14] and it may be concluded that its mass is small.

Taking into account all these difficulties let us consider some other possible forces of a non-gravitational character. In this connexion we may indicate the forces under the influence of which the prominences of the surge type are moving. For example, for the surge that was observed over the chromospheric flare of 8 May 1951 [15], decelerations were registered which were very large compared with gravitation. The existence of such forces which act in the direction towards the centre of the star and which exceed the force of gravitation, must, therefore, be considered as a fact, which follows directly from observations! Hence the necessity of introducing these forces (obviously of an electro-magnetic nature) cannot be considered as a difficulty of hypothesis B. Moreover, the phenomenon of surges ejected and then returned to the Sun with accelerations exceeding gravitation is very similar to the phenomenon of expansion of a nova. The only difference here is that the separate condensations forming the photosphere are moving radially in *all* directions.

It is of interest to note that there exists another similarity between the outbursts of novae and the chromospheric flares. It is known that the appearance of a chromospheric flare on the Sun is accompanied in certain cases by a growth in the intensity of cosmic rays. On the other hand, there are some reasons to think [16] that the envelopes ejected from the supernovae and from the novae contain particles of very high energies. Thus, in both cases we have simultaneously: (a) a sharply non-stationary process, accompanied by ejection and consequent return of matter, (b) an appearance of forces that exceed gravitation, (c) the formation of high energy particles. Obviously, such an extremely complicated non-stationary process characterizes not only the novae and the chromospheric flares, but also many other phenomena in the atmospheres of the non-stationary stars.

The idea that the rise of the electromagnetic forces inside the expanding nova, which are directed to its centre, may be connected with the phenomena of the outburst itself, is supported by the following considerations. It is known that the outburst of a nova is accompanied by ejection of separate gaseous condensations that move radially with different velocities. Ionized gas clouds moving with high velocities pass through clouds with low velocities. This causes* the formation of electric currents, the directions of which are radial. These currents will be accompanied by magnetic fields with components perpendicular to the radius.

The main source of the magnetic fields must be the inner parts of the expanding nova, where all the processes connected with the explosion proceed most intensively. These magnetic fields will cause the retardation of the conducting gases of the photosphere and of the reversing layer. That the magnetic fields may be large, even at distances of several hundred stellar radii from the surface of the star, is evident in the case of AG Pegasi (see reference 17).

It is necessary to point out here that the magnetic fields thus formed (as well as the general magnetic field of the star which possibly existed before the outburst (see below)), must be greatly strengthened by further chaotic turbulent movements of the conducting gases. The 'interlacing' of the magnetic lines of forces will make the star resemble a ball of elastic threads and this will also cause retardation of radially moving gases.

It is possible that the initial general magnetic field of the star plays the principal role in the problem of retardation. Indeed, if before the outburst a nova possesses a general magnetic field, the retardation of ejected ionized matter will be the least in the polar directions, where magnetic lines of force are approximately radial. Accordingly, the quantity of ejected matter in these directions will be the greatest. This is in agreement with the fact that in many cases the main mass of matter is ejected during the outburst in two diametrically opposite directions. We have in mind Baade's polar caps for Nova Aql 1918, the two condensations in the nebula of Nova Her 1934, and generally the so-called 'doubling' of novae. Finally, we must indicate the symmetrical disposition of the two maxima in the emission bands in the spectra of the majority of novae. Table 1 illustrates this fact.

However, it must be stated at the same time that in spite of the very complex character of the phenomena of ejection before light maximum the expansion of each new star proceeds in *all* directions. In certain direc-

* Because of the different masses of electrons and ions and their different cross-sections.

tions, however, the quantity of ejected matter is the greatest. The fact that for all the seven new stars that were 'caught' before light maximum, a replacement of the pre-maximum spectrum by the principal spectrum took place speaks in favour of this picture. Therefore the general concepts of the mechanism of expansion considered above must be retained.

Table 1. *Data on the Presence of Two Maxima in the Emission Bands of Novae*

Nova Per 1901	absent	Nova Mon 1939	absent
Nova Lac 1911	present	Nova Pup 1942	present?*
Nova Gem 1912	present	Nova Aql 1943	present
Nova Aql 1918	present	Nova Aql 1945	absent
Nova Cyg 1920	present	Nova Sgr 1947	absent
Nova Pic 1925	present	Nova Cyg 1948	absent
Nova Her 1934	present	Nova Ser 1948	present
Nova Sgr 1936	present	Nova Sco 1950 (1)	present
Nova Aql 1936 (1)	present	Nova Lac 1950	absent
Nova Lac 1936	present	η Car	present

* Present in He I, He II.

It is of interest to point out the following fact. Let us assume that the radio-frequency radiation from the envelopes ejected by super-novae is actually connected with the emission of radiation of the relativistic electrons in magnetic fields[16]. In this case we can consider that these envelopes in their *present* state are characterized by the presence of magnetic fields. On the other hand, magnetic fields in these envelopes must change with time extremely slowly. Therefore, magnetic fields must be present in these envelopes at the moment of their *detachment* from super-novae. It is possible that the same is true for novae.

The fact that during the outbursts of novae some considerable inward forces must arise is confirmed by a number of subsidiary considerations. In particular, it is of interest to note that in Fig. 3 even the dotted line *a–a* corresponds to velocities exceeding the parabolic. The impression is that practically all the mass of the expanded envelope of Nova Aql 1918, including the region with $\tau > 1$, should have left the star, but this seems hardly admissible.

We have already pointed out that the source of matter which forms the diffuse-enhanced spectrum is localized in the external layers of the contracting nova. The source of the Orion spectrum is also located somewhere not very deep under the photosphere, but not in the main body of the star. Therefore it may be accepted that the processes of continuous ejection of matter after light maximum are connected with the internal instability of the outer layers of the contracting nova, but not with the continuous

67

'explosions' in the main body of the star. It may be suggested that the source of such non-stability is the process of contraction of the star (after its light maximum) which leads to a transformation of the contraction energy into thermal energy and into energies of some other types. However, forces which greatly exceed gravitation at $m = m_\odot$ are required in order to provide a sufficiently rapid production of energy of contraction. It should be taken into account that sometimes the diffuse-enhanced spectrum of novae appears a few hours after light maximum.

The fact that after light maximum no jump of intensity at the limit of the Balmer series is observed [2] speaks in favour of an extreme non-stability of the outer layers of novae at this time (a Balmer jump due to recombination appears much later, as in the case of Nova Lac 1936, and is connected with the phenomena of fluorescence). This, as was shown by the author [18], may obviously be explained by the fact that the outer layers of a nova after its light maximum are in a relatively isothermic state, which requires random gas movements with supersonic velocities. The formation of emission bands of the principal spectrum [10, 19] is evidently connected with these processes. As we have seen, the temperature of the star up to the time t_e does not increase, but decreases slightly. Therefore, fluorescence of a thermal character must be excluded. It may be expected that the energy arising in the extended envelope as a result of atomic collisions* (and other processes) is absorbed by the principal envelope in its wide (ultra-violet) absorption lines. Following this, a re-emission of energy takes place in other lines as well.

Furthermore, it is of interest to point out that although the source of formation of the diffuse-enhanced and Orion spectra is the instability of the corresponding regions of the nova, the particular mechanism of the ejection of matter is similar in this case [7] to the ejection of prominences from the Sun. This emphasizes again the considerable role of the electromagnetic forces in the phenomenon of novae.

The author's computations show [7] also that atoms forming the diffuse-enhanced spectrum must be accelerated by selective radiation pressure in Lα. The intensity of Lα, estimated from the intensity of the observed Hα line, is expected to be very high.

A very important problem in the physics of novae is the calculation of the mass m of the envelopes ejected by novae. This problem was first considered by V. Ambartsumian and N. Kozyrev [20].

The masses of envelopes are usually estimated from the intensity of emission bands created by the principal envelope. Recently (see above)

* Owing to collisions between gaseous clouds moving with different turbulent velocities.

68

this method has been used by I. Kopylov[13]. He showed that there is a linear dependence between log m and the absolute visual brightness of the nova before its outburst. The brighter the star is before its outburst, the greater is log m.

Another method[21] of estimating the mass m is based on the observed accelerations of envelopes after light maximum. These accelerations $j > 0$ may be connected with the following forces: (a) the general and selective radiation pressure, or (b) the 'corpuscular' pressure. Atoms producing the diffuse-enhanced and Orion spectra have velocities larger than the velocity of the principal envelope. Overtaking the latter, they continuously communicate to it a part of their momentum.

Application of this method[21] gave values of m that are of the same order as the masses m calculated by other methods.

Let us consider briefly the problem of the origin of the outbursts of novae. It is clear that the novae possess some properties which distinguish them from other stars. It is possible that this is connected[22] with anomalies in the chemical composition of novae. The existing data indicate that many novae are characterized by a high content of the light elements O, C, N. Quantitative investigations are needed here.

The investigation of the outbursts of novae is of great importance from the dynamical and energetic point of view. The process of cooling of a nova during its expansion is especially interesting. The fact that the outer layers of the expanding nova emit radiant energy plays a very important part. This process, connected with the diffusion of light quanta in the envelope of a nova, has been considered recently by V. Sobolev[23].

REFERENCES

[1] A. Beer, *M.N.* **97**, 231 (1937).
[2] E. R. Mustel, *A.J. U.S.S.R.* **22**, 65, 185 (1945).
[3] E. R. Mustel, *A.J. U.S.S.R.* **23**, 289 (1946).
[4] D. McLaughlin, *Michigan Obs. Publ.* **8**, No. 12; *Pop. Astr.* **52**, 109 (1944).
[5] E. R. Mustel, *Publ. Crim. Astrophys. Obs.* **4**, 152 (1949).
[6] E. R. Mustel, *Publ. Crim. Astrophys. Obs.* **1**, part 2, 91 (1948).
[7] E. R. Mustel, *A.J. U.S.S.R.* **24**, 97, 155 (1947).
[8] E. R. Mustel, *Comptes rendus de l'Acad. d. Sci. de l'U.R.S.S.* **29**, 296, 365 (1940).
[9] E. R. Mustel, *Publ. Crim. Astrophys. Obs.* **4**, 23 (1949).
[10] E. R. Mustel, *Vistas in Astronomy*, v. **2**, Pergamon Press, London (1956).
[11] E. R. Mustel, *A.J. U.S.S.R.* **24**, 280 (1947).
[12] R. F. Sanford, *Ap.J.* **109**, 81 (1949).
[13] I. M. Kopylov, *Publ. Crim. Astrophys. Obs.* **10**, 200 (1953).
[14] M. Walker, *Publ. A.S.P.* **66**, 230 (1954).
[15] T. Bartlett, B. Witte, and W. Roberts, *Ap.J.* **117**, 292 (1953).

[16] I. S. Shklovsky, *A.J. U.S.S.R.* **30**, 577 (1953); see also *Les processus nucléaires dans les astres*, Liège, 1954, p. 515.
[17] G. Burbidge and E. Burbidge, *Ap.J.* **120**, 76 (1954).
[18] E. R. Mustel, *Publ. Crim. Astrophys. Obs.* **7**, 118 (1951).
[19] E. R. Mustel, *A.J. U.S.S.R.* **25**, 156 (1948).
[20] V. A. Ambartsumian and N. A. Kozyrev, *Zs. f. Ap.* **7**, 320 (1933).
[21] E. R. Mustel, *Comm. Sternberg Astronomical Inst.* No. 41 (1950).
[22] E. R. Mustel, *Publ. Crim. Astrophys. Obs.* **6**, 144 (1951).
[23] V. V. Sobolev, *A.J. U.S.S.R.* **31**, 15 (1954).

10. SUB-SYSTEM OF NOVAE

I. M. KOPYLOV

Crimean Astrophysical Observatory, Pochtovoje, Crimea, U.S.S.R.

In order to discover the cause of the outbursts of novae, a study of the physical processes taking place in novae during such explosions is not sufficient. A number of other, and no less important, properties and peculiarities of novae must also be investigated. Such properties are, for instance: the absolute magnitudes at light maximum and minimum, the temperatures, the light curves and their variations, the position of the novae on the Hertzsprung-Russell diagram, their spatial distribution in the Galaxy, and the mutual relations between the novae and other objects of similar nature (such as the recurrent novae, super-novae, nova-like stars, and the nuclei of planetary nebulae). A hypothesis of the origin of the outbursts of novae should be founded upon all these data as well as the other properties of novae.

A knowledge of the absolute magnitudes of novae at light maximum ($M_{max.}$) is required for the solution of a number of questions on the physical properties of novae.

The most reliable method for a determination of the absolute magnitudes of novae (from the expansion of the ejected envelopes) permitted McLaughlin in 1945 to define (with the aid of some other methods and with allowance for space absorption) the relationship between the absolute magnitude of a nova at light maximum, $M_{max.}$, and the time t_3 in days of the decline of the light of the nova for 3 magnitudes after the light maximum. The dependence ($M_{max.}$, t_3) (see Fig. 1) can be expressed by the equation:

$$M_{max.} = -13^{m}\cdot7 + 3^{m}\cdot6 \log t_3. \tag{1}$$
$$\pm 0\cdot1 \quad \pm 0\cdot2$$

The absolute magnitude of a nova at light maximum can thus be determined according to the value of t_3 from the light curve. Equation (1) affords a most simple method for the determination of the absolute magnitudes, and, consequently, of the distances of the novae.

A study of the light curves of novae permits one to establish the following two relations:

(a) Immediately after maximum the light of a nova declines according to the law:

$$M_i = -13^{m}\cdot7 + 3^{m}\cdot5 \log t_3^i, \tag{2}$$

where M_i is the absolute magnitude of the nova at a given moment (at the moment of light maximum in particular), and t_3^i is the time of the subsequent decline of its light for 3 magnitudes.

(*b*) The duration of the total decline of the light of a nova from maximum to minimum is determined by the luminosity of the nova at maximum.

These two peculiarities of the light curves of novae express, in fact, the same important general property of the novae: the rate of the subsequent

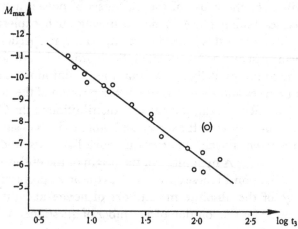

Fig. 1. The dependence of the absolute magnitude of novae at maximum light upon the time (in days) required for the light to decline 3 magnitudes below maximum.

decline of the light of a star is determined by the value of its luminosity at a given moment. Similar relationships between phenomena and the rates of their evolution can also be observed in some other astrophysical processes. For example, they appear in the outbursts of super-novae, the development of the chromospheric flares on the Sun, and so on.

The relation (2) permits one to determine the absolute magnitude of a nova from any part of the light curve where the luminosity of the star decreases not less than 3 magnitudes. This permits one to extend considerably the data regarding the absolute magnitudes and distances of novae, using only their light curves.

The luminosity function of novae at light maximum obtained in the first approximation has the following parameters: a mean value of $M_{max.} = -7^{m}\cdot5$, extreme values of $-2^{m}\cdot0$ and $-11^{m}\cdot0$, and a dispersion of $\pm2^{m}\cdot5$.

The luminosity function of novae in their normal state has these parameters: a mean value of $+3^{m}\cdot1$, extreme values of $-1^{m}\cdot5$ and $+7^{m}\cdot5$, and a dispersion of $\pm2^{m}\cdot5$. The parameters of the distribution of novae according to the values of their amplitudes are: a mean value of $10^{m}\cdot 4$, extreme values of $6^{m}\cdot 0$ and $15^{m}\cdot 0$, and a dispersion of $\pm 3^{m}\cdot 5$.

A comparison with super-novae. The faintest super-novae (about $M = -13$ at light maximum) are approximately 2 magnitudes brighter than the most luminous novae (of about $M = -11$). The absence of exploding stars of intermediate absolute magnitudes (from -11 to -13^{m}) is obviously real. The mean absolute magnitude of super-novae at light maximum is about $-15^{m}\cdot 7$, and of typical novae only about $-7^{m}\cdot 5$.

The amplitudes of the super-novae are not less than 21 magnitudes. The largest amplitudes of the novae can, apparently, reach but 15 to 17 magnitudes.

For super-novae as well as for novae, a relation $(M_{max.}, t_3)$ holds true. This relation for super-novae is expressed by the equation

$$M_{max.} = -31^{m}\cdot 9 + 9^{m}\cdot 5 \log t_3. \tag{3}$$
$$\quad\;\; \pm 0\cdot 3 \quad\; \pm 0\cdot 2$$

It is easy to see that the relation $(M_{max.}, t_3)$ for super-novae is not an extension of the similar relation (1) for typical novae. The spatial distribution of novae and super-novae in the Galaxy is also altogether different. The super-novae form a flat sub-system, while the typical novae form an intermediate sub-system. Thus the novae and super-novae form two independent groups of stars, clearly differing both in the distribution of their amplitudes and absolute magnitudes at maximum, as well as in their distribution in the Galaxy.

A comparison with the recurrent novae and U Geminorum type stars. The novae were previously considered to form an extension of the recurrent novae towards the region of large amplitudes and large periods. This assumption was based upon the 'amplitude-cycle' relation, which was established for the nova-like U Geminorum type stars. Recurrent novae were assumed to satisfy this relation, as well as were the typical novae. But the accumulation of observational data concerning stars of these types shows quite clearly intrinsic differences in the physical properties, as well as in the mechanisms of the outbursts, between the nova-like stars and the recurrent novae.

It was found also that the 'amplitude-cycle' relation is different for both groups of stars. The 'amplitude-cycle' relation for nova-like stars is

$$\log P = -0\cdot 38 + 0\cdot 60\, A, \tag{4}$$

and for the recurrent novae is

$$\log P = +2\cdot 83 + 0\cdot 14\, A, \tag{5}$$

where A is the amplitude and P is the period expressed in days (Fig. 2).

A comparison of the amplitudes of the typical and recurrent novae shows that the novae do not satisfy relation (5), established for the recurrent novae, because the typical novae do not explode a second time, in spite of having approximately the same amplitudes as do the recurrent novae. Since repeated outbursts of novae have never been observed, the question of the intervals between such bursts remains open. It is quite possible that outbursts of typical novae occur only once in the course of their lifetime. It should be remembered in this connexion that all the statistics concerning

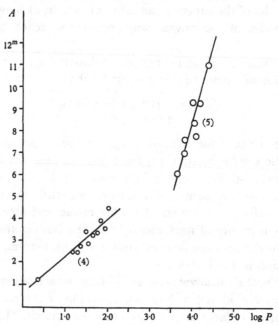

Fig. 2. The dependence of amplitude upon period for the nova-like U Geminorum variables (the straight line is that of equation (4)) and for the recurrent novae (the line is that of equation (5)).

the numbers of novae in the Galaxy are based upon their frequency in the present stage of the Galaxy.

The recurrent and typical novae form sub-systems of different flatness in the Galaxy.

The data quoted, together with additional information, permit one to suggest that there is no genetic connexion between the typical and recurrent novae and that they represent two altogether different types of stars.

A comparison of novae with the nuclei of planetary nebulae. A common feature of the novae and the nuclei of the planetary nebulae is the presence of expanding envelopes. The envelopes of novae have smaller masses (10^{-3} to 10^{-4} m$\odot$) and the velocity of their expansion is larger ($v \sim 1000$ km./sec.)

74

than in the case of planetary nebulae ($m \sim 10^{-1}$ to 10^{-2} $m_\odot$, and $v \sim 10$ km./sec.).

Novae (in their normal state) and the nuclei of the planetary nebulae occupy about the same region in the spectrum-luminosity diagram, namely that of high temperatures and moderate luminosities. The luminosity functions of the novae and of the nuclei of planetary nebulae show a close similarity: $\overline{M}_{min.} = +3 \cdot 1$ for novae and $\overline{M} = +2 \cdot 5$ for the nuclei. The extreme values of the absolute magnitudes are -2^m and $+8^m$ for novae, and -2^m and $+7^m$ for the nuclei of planetary nebulae. Thus, they are practically coincident. The spectra and, consequently, the temperature of the novae (in their normal state) and of the nuclei of planetary nebulae are also practically the same.

The apparent distribution in the sky of novae and planetary nuclei is rather similar. Stars of these two types form intermediate sub-systems with practically identical parameters. Their mean spatial velocities are also similar. From such a comparison of the novae with stars of other types, it follows that the nuclei of planetary nebulae show the closest similarity to the novae.

On the basis of these results, the following may be said about the origin and evolution of novae. In accordance with the ideas of Vorontsov-Velyaminov, who supposes an evolution of stars along the blue-white sequence, it appears that the novae represent an intermediate stage between the recurrent novae and the white dwarfs.

The differences that have been pointed out between these stars make the transfer of the recurrent novae into the typical novae highly improbable, if the following two circumstances in particular are considered.

(a) The novae do not satisfy the amplitude-cycle relation for the recurrent novae, and (b), they form a sub-system in the Galaxy totally different from that of the recurrent novae.

These two types of stars are, probably, of quite different origin and age.

According to rather rough estimates $\overline{M}_{min.} = +2 \cdot 9$ for the recurrent novae, $\overline{M}_{min.} = +3 \cdot 1$ for the typical novae and $\overline{M} = +2 \cdot 5$ for the nuclei of planetary nebulae. All these three types of stars are extremely hot objects. Thus, stars of these types, which eject (in some way or other) discrete envelopes, are located in the same region of the luminosity-spectrum diagram. It may be admitted that the evolution of such stars follow, in general, parallel courses, but nothing conclusive can at present be said regarding the final stage of evolution of novae. We are not inclined to exclude completely the possibility that the white (or blue) dwarfs are the final stage of the development of novae, since the parameters of their sub-systems are very similar. However, the absence of blue and white stars in the range of absolute magnitude between $+7 \cdot 5$ and $+9^{m} \cdot 5$ decreases this possibility immensely, such a gap being quite incomprehensible in terms of the hypothesis of the evolution of novae into white dwarfs.

From all the foregoing results it follows, in any case, that the interpretation of the blue-white sequence as an evolutionary track of hot giant stars should be altogether rejected.

A table containing the parameters of the spatial distribution in the Galaxy of novae and the objects that are to some extent related with novae is given below:

Galactic Distribution Parameters for Super-novae, Novae, Planetary Nuclei, and Similar Objects

Sub-system	$\dfrac{\partial \log D}{\partial R}$	$\dfrac{\partial \log D}{\partial z}$	β (parsecs)
Super-novae	$-0 \cdot 15$	$-5 \cdot 2$	83
Nuclei of planetary nebulae	$-0 \cdot 21$	$-2 \cdot 20$	197
Typical novae	$-0 \cdot 22$	$-2 \cdot 39$	182
White dwarfs	$-0 \cdot 23$	$-2 \cdot 7$	160
Recurrent novae	$-0 \cdot 25$	$-0 \cdot 9$	480
Nova-like U Geminorum type stars	$-0 \cdot 27$	$(-0 \cdot 17)$	2600

11. ON THE CAUSES OF STELLAR BURSTS

L. E. GUREVITCH AND A. I. LEBEDINSKY

Physico-Technical Institute, Leningrad, U.S.S.R.;
University, Moscow, U.S.S.R.

The phenomena taking place as a result of a nova outburst indicate that a shock wave is propagated to the surface of the star. It may be the result of an explosion in the interior of the star. The hypothesis that peripheral thermo-nuclear explosions cause the outbursts of novae, and perhaps also those of nova-like stars, was advanced in 1947[1]. By 'peripheral' we mean some spherical layer which we shall name the A layer.

A thermo-nuclear explosion in the central parts of the star is impossible under ordinary conditions because the star possesses the property of 'self-adjustment', which prevents an explosion in its interior. If as a result of a gradual increase in the activity of the central energy sources, the excessive heat has sufficient time to become distributed uniformly throughout the star, then, according to the virial theorem, an expansion of the star results. This leads not to a heating, but to a cooling of the central parts of the star. As will be explained later on, a peripheral explosion is not only possible but is even inevitable in certain evolutionary stages of some stars.

The hypothesis of a peripheral explosion is confirmed by the observational data on the asymmetry of envelopes ejected by the novae. Such asymmetry was clearly present in the cases of Nova Pictoris 1925 and Nova Herculis 1934, which resembled multiple stars some time after the outbursts. A central explosion in a spherically symmetric star develops without any violation of the central symmetry and must therefore cause the ejection of a spherical envelope. Conversely, as the result of a peripheral explosion the star becomes asymmetric, and an asymmetric envelope is ejected. This occurs owing to the fact that the explosion, which took place at some point of the A layer, propagates in the form of a detonation wave throughout that layer during a period of time comparable with the time required for the passage of the shock wave from the A layer to the surface of the star.

The velocities with which the envelopes are ejected by novae exceed the value appropriate for the region of the star where thermo-nuclear explosions, due to hydrogen being converted into helium, would be expected.

This circumstance has been explained both qualitatively[2] and quantitatively[3] by the increase of W, the velocity of the shock wave propagated in a sphere of outwardly decreasing density, when the front of the wave advances from the interior to the peripheral parts of the star. In a spherically symmetrical star $W^3 r^2 \rho = $ constant, where ρ is the density previous to the explosion at a distance r from the centre of the star.

A peripheral thermo-nuclear explosion can take place, as was shown by us[4], when a star reaches a stage of its evolution when, as a consequence of hydrogen exhaustion in its central part, an isothermal region of such large dimensions is formed that the star passes from the condition where nuclear reactions provide its main source of energy, to the regime of gravitational contraction. Evolution of a star in this stage consists of contraction, accompanied by a gradual heating of its whole mass including its outer parts where some hydrogen remains.

When the transition of the star from the regime of nuclear energy sources to the regime of gravitational contraction is taking place, the star passes into the condition of instability, in which outbursts are possible. To explain this, let us investigate the most simple case (in the mathematical sense), where the outer parts where hydrogen is still present have become thin and their lower boundary lies at a depth z_A, which is small compared to the radius of the star R. In this case, the temperature T_A at a depth z_A is determined by the formula

$$T_A \propto z_A \propto \left(\frac{M_e}{R^2}\right)^{4/17} (gF)^{2/17}, \qquad (1)$$

where M_e is the mass of the outer region, while g and F are the acceleration of gravity and the flux density of radiation on the surface of the star.

For a random increase of the energy source, F will increase along with it. This will also increase T_A according to (1). An increase of T_A will also intensify the energy output of the nuclear reactions in the A layer, which will in turn cause further growth of F and T_A. For a random decrease of F, the depth z_A, and at the same time also the temperature T_A will decrease, and this will lead to a decrease of the nuclear reactions and to a further fall of F. Such a state is unstable, as it may be proved, and will inevitably lead to an explosion.

Previous to the onset of non-stability, the depth z_A is large and relation (1) is incorrect. An occasional change of the flux of energy will cause a change of the temperature in the opposite direction. This will again re-establish the normal flow of thermo-nuclear energy. It is quite probable that this non-stability was formally discovered, without its physical significance being noted, by Chandrasekhar and Henrich[5]. They suggested that when the isothermal region of a star reaches a definite critical dimension the equilibrium equations become incompatible.

A slow evolutionary heating of the star causes a gradual temperature increase in the A layer, which continues up to some critical value T_k. At this stage, the radiative transport of energy cannot keep up with the output of the thermo-nuclear sources and a continuously increasing heating of the A layer begins. Over a period of time, which we shall name the induction period, the temperature at the lower boundary of the A layer will grow from the value of T_k to a value of T_i. At this stage, the thermo-nuclear sources in the A layer will be so intense that a discontinuity in the pressure will be formed. The shock wave, which carries a part of the energy to the surface, causes a strong heating of the gas masses lying above the A layer.

An examination of the kinetics of the outburst process[4] permits the following estimates: T_k is about 2×10^7 degrees, and T_i is about 8×10^7 degrees. The induction period in the case of a typical nova continues for several thousands of years, and in the case of outbursts of super-novae, for millions of years. But only immediately before the outburst—approximately a week before in the case of novae—the heat emission in an A layer will increase to such an extent that a super-adiabatic temperature gradient will be reached at the outer boundary of the A layer. This cannot cause a mixing of matter, because a considerably longer time is necessary for convection to develop, as may be shown.

A hydrogen explosion will rapidly be halted or, as we express it, become degenerated, so that only a small portion of the hydrogen has time to become converted into helium. The cause of the degeneration of the explosion lies in the properties of nuclear reactions. Out of the two known reactions for the transformation of hydrogen into helium, only the carbon cycle leads to an explosion. The other reaction, the proton-proton cycle, cannot cause an explosion owing to its slowness. As was shown by us[4], the output of the thermo-nuclear source in the A layer must be comparable to the luminosity of the star at the time the critical temperature is reached. At the moment the explosion starts, it must exceed that luminosity by a number of orders of magnitude. The proton-proton reaction cannot produce such emission. The Bethe reaction, however, involves β-decays that take place during times of the order of some hundred seconds[6]:

$$N^{13} \rightarrow C^{13} + e^+,$$

the half-life being $9 \cdot 93 \pm 0 \cdot 03$ min., and

$$O^{15} \rightarrow N^{15} + e^+,$$

79

the half-life of which is 115 ± 1 sec. A possible path of the reaction

$$N^{13} + p \to O^{14} \to N^{14} + e^{+}$$

also involves β-decay, the half-life equalling 76.5 ± 0.5 sec. The velocity of the Bethe cycle at high temperatures is limited by the above processes. Therefore, in the course of time of one β-decay, every carbon or nitrogen nucleus may capture on the average only one proton. During this time a part x_c/x_H of the hydrogen abundance is converted into helium. Here, x_H and x_c are the atomic concentrations of hydrogen and carbon-nitrogen. At x_H about 1 and the ordinary value of x_c about 10^{-3}, this fraction equals 10^{-3}, so that the whole hydrogen abundance may be spent in a day. But the gases of the outer region, heated up to many tens of millions of degrees, cannot be at rest for a day. The outer region will inevitably reach during a time of the order of some tens of seconds a state of rapid expansion, accompanied, generally speaking, by disorderly convective motions. A considerable cooling is thus produced, and the explosion is stopped. Therefore, the amount of hydrogen that will have time to be transformed into helium during the outburst is of the order of the abundance of carbon atoms, namely 10^{-3}. While only a small amount of hydrogen burns away during the explosion, the same conditions are re-established afterwards. The explosion must, therefore, repeat itself after a certain lapse of time.

The sum of the observational facts makes it rather probable that outbursts of every nova are actually recurrent. The intervals of time between outbursts must be of the order of the time required for a pre-explosional heating. A strong outburst, causing an intense turbulent motion, cools the stellar gas after the explosion, so that the gas almost reaches temperature T_k. The intervals of time between such bursts are, therefore, comparable to the induction time. Minor outbursts cause a lesser cooling and the intervals between them are shorter. Thus, a possibility appears of explaining the correlation between the frequency of outbursts and their amplitudes.

The energy scale of an outburst is determined by the mass of the outer part of the star containing hydrogen, namely by the mass M_e. This mass has previous to the induction period a temperature less than T_k. An evaluation of this mass leads to the equation

$$M_e = 6.5 \frac{R^4}{\sqrt{mL}} \left(\frac{T_k}{10^7} \right)^{17/4},$$

where m is the mass, L the luminosity, and R the radius of the star before its outburst, expressed in solar units. The explosion does not involve the hydrogen as a whole, but only the innermost layers with masses equalling o·16 M_e, as may be seen from calculations. Consequently, the energy of the explosion is

$$Q = \frac{q x_c R^4}{\sqrt{mL}} \left(\frac{T_k}{10^7}\right)^{17/4}, \qquad (2)$$

where q is the amount of energy emitted when a mass of hydrogen equal to the mass of the Sun is converted into helium.

As is seen from equation (2), outbursts on the energy scale of novae may occur only in the case of sufficiently dense dwarfs that possess masses and luminosities comparable to the Sun, and a radius of the order of less than 0·1 $R_\odot$. According to Walker's data[7], Nova Herculis 1934 is a dwarf of that kind.

Outbursts on the scale of super-novae can take place during the first explosions, when the mass M_e constitutes a major part of the star. If x_c is greater than 2–3 %, then a hydrogen explosion may call forth a detonation of the reaction when helium converts into carbon. In this reaction β-decay is not involved, and it may therefore cause an explosion of the main mass of the star. At x_c less than 0·005, the temperature does not reach the value $T_i = 8 \times 10^7$ degrees. Expansion of the A layer therefore proceeds with a sub-sonic velocity and the shock wave is formed in the higher layers.

Along with the stellar outbursts resulting from the preceding internal causes, it has been shown by one of the authors[8] that outbursts caused by external factors and particularly by the accretion of interstellar gases containing deuterium are also possible.

As we have already seen, all the processes of the Bethe cycle proceed with extremely large velocities (except the processes involving β-decay that lead to the formation of C^{13} and N^{15}) when extremely high temperatures are attained in the A layer during the course of the explosion. The increase of thermo-nuclear emission is stopped and the explosion degenerates when a considerable part of the stable nuclei of C^{12}, C^{13}, N^{14} and N^{15} are converted into radioactive nuclei N^{13}, O^{15} and O^{14}, which later pass into C^{13}, N^{15} and N^{14} by β-decay. Therefore, as a result of outbursts, C^{12} and N^{14} must become converted into C^{13} and N^{15} and, consequently, the ratios $C^{13}:C^{12}$ and $N^{15}:N^{14}$ must be anomalously large in the atmosphere of stars in which explosions have taken place.

In some cool stars an anomalously high abundance of the isotope C^{13} has been discovered[9]. The observed value of the ratio $C^{13}:C^{12}$ reaches unity. From our point of view, this may be a consequence of the fact that these stars have experienced outbursts in the past. The possibility is not excluded that the ratio $N^{15}:N^{14}$ is also anomalously large in some of these stars. It should be extremely interesting to check this supposition by means of observations.

REFERENCES

[1] L. E. Gurevitch and A. I. Lebedinsky, *Comptes rendus de l'Acad. d. Sci. de l'U.R.S.S.* **56**, 25, 137 (1947); *Jour. Exp. Techn. Phys.* **17**, 792 (1947).

[2] A. I. Lebedinsky, *Scientific Papers of Leningrad University*, No. 6, 3, 1946; *Annals of the Scientific Session of the Leningrad University devoted to its Jubileum, section of mathematical Sciences*, **12**, 1947.

[3] A. I. Lebedinsky, *A.J. U.S.S.R.* **29**, 135 (1952).

[4] L. E. Gurevitch and A. I. Lebedinsky, *Publications of the Fourth Cosmogonical Conference*, p. 143, Moscow, 1955.

[5] S. Chandrasekhar and L. R. Henrich, *Ap.J.* **94**, 525 (1941).

[6] I. P. Selinov, *Atomic Nuclei and Nuclear Transformations*, v. **1**, p. 24, Moscow, 1951.

[7] M. F. Walker, *Publ. A.S.P.* **66**, 230 (1954).

[8] L. E. Gurevitch, *Problems of Cosmogony*, v. **2**, p. 150, Moscow, 1951.

[9] *Astrophysics*, ed. by J. A. Hynek, p. 153, New York, 1951.

12. ON THE NATURE OF PLANETARY NEBULAE

I. S. SHKLOVSKY
Sternberg Astronomical Institute, Moscow, U.S.S.R.

The physical state of the planetary nebulae has been investigated sufficiently well, but the main question of their origin and evolution still remains open.

For the solution of this most significant problem, it is very important in our opinion to analyse the well-known fact that the planetary nebulae undergo an unlimited expansion, and a consequent dissipation. According to observations, the velocities of expansion of the planetary nebulae equal several tens of kilometres per second. The same velocity should also be expected when hot and comparatively dense gases expand into a vacuum.

Two principal conclusions may be drawn from the fact of infinite expansion of planetary nebulae: (a) the optical thickness of each planetary nebula decreases continuously until it becomes less than unity. (b) The luminosity of an optically thin planetary nebula decreases with the evolution of the nebula approximately as r^{-3}, which is proportional to t^{-3}, where r is the radius of the nebula and t is the time. The surface brightness decreases approximately as r^{-5}, which is proportional to t^{-5}.

The brightness of the majority of planetary nebulae will, over several thousands of years, decrease to such an extent that they will no longer be observable. It follows then that the process of continuous formation of planetary nebulae is going on in the Galaxy. The number of planetary nebulae formed every year in our Galaxy is of the order of unity.

No adequate methods exist at the present time for the determination of the distances of the planetary nebulae. It can be shown that the well-known methods of Vorontsov-Velyaminov, Berman, and Camm cannot be considered as correct. For example, the main assumption in Vorontsov-Velyaminov's method that all planetary nebulae have the same luminosity is not correct, because as a result of their expansion, the luminosities of the planetary nebulae decrease indefinitely. The young, star-like planetary nebulae must possess the highest luminosities and very high surface brightnesses. The luminosities of the old, greatly expanded objects with low surface brightnesses must be the lowest.

In accordance with these ideas, we have developed a new method for the determination of the distances of the planetary nebulae, based upon physical reasoning. From simple reasoning, it follows that the distance of an optically thin planetary nebula is

$$R \propto \frac{\mathbf{m}^{0.4}}{\phi I^{0.2}},$$

where $\mathbf{m}$ is the mass, I the surface brightness, and ϕ the angular dimensions. Estimates of the distance depend but very little upon the distribution of the surface brightness, because R depends to a very small extent upon I. A new catalogue of planetary distances has been computed. The masses of planetary nebulae have been considered for this purpose to be equal. This assumption cannot influence the results seriously, because R depends but slightly upon $\mathbf{m}$. The zero-point of the new system of distances was determined from an analysis of the known proper motions and radial velocities, the interstellar absorption being taken into account. The newly derived distances differ greatly from the previously accepted values. Thus, for instance, NGC 7293 has a distance of 50 parsecs in the new catalogue, while according to Vorontsov-Velyaminov it is 250 parsecs, and according to Berman it equals 1050 parsecs.

It should be mentioned that Van Maanen determined a parallax of $0''038 \pm 0''008$ for the nucleus of NGC 7293; this fact was, however, not taken into account in the present analysis. It may be shown that if NGC 7293 had a distance exceeding 250 parsecs, its mass should be larger than 10 $\mathbf{m}_\odot$. Such a value is hardly acceptable. For $R = 50$ parsecs, the mass would be about 0.2 $\mathbf{m}_\odot$.

An analysis of the new system of distances permits one to draw some conclusions about the absolute magnitudes of the nuclei of planetary nebulae. The nebulae themselves show an extremely large scatter, ranging from $M = -0.5$ to $+10$. The absolute magnitudes of a number of nuclei of planetary nebulae range from $M = +5$ to $+10$. As the temperatures of the nuclei are extremely high, it may easily be shown that these stars possess mean densities similar to those of white dwarfs. Consequently, the nuclei of a number of planetary nebulae are 'over-heated' white dwarfs.

We believe such a conclusion is of particular importance in a consideration of the planetary nebulae. After some tens of thousands of years, the dimensions of the nebula NGC 7293 will increase several times. Its brightness also must diminish by a factor of several hundred, and the nebula will become invisible. Only the nucleus will remain as an extremely hot white dwarf. The nucleus will cool and change gradually into a normal white dwarf. The absolute magnitudes of the majority of nuclei of planetary nebulae considerably exceed the absolute magnitudes of white dwarfs. This fact may indicate that along with a rapid evolution of planetary nebulae, there also takes place a rather rapid (but not an explosive) evolution of their nuclei, the luminosity and temperature of the nuclei thereby decreasing. This possibility may explain the well-known fact that the spectral characteristics of the various nuclei correspond to different morphological types of the associated nebulae. The genetic connexion between planetary nebulae and white dwarfs suggests that some stars at a definite stage of their evolution detach a shell with zero velocity in the process of transformation into white dwarfs. This process would go on for several tens of thousands of years.

Great theoretical difficulties arise if such an interpretation is accepted, but they do not seem insurmountable to us.

It is important to note that the creation of the planetary nebulae from stars is not a recurrent process. This conclusion follows from the value of the mass of planetary nebulae (**m** about 0.2 **m**⊙).

Several billions of planetary nebulae have formed and then decayed during the life-period of the Galaxy. It has recently been estimated that the number of white dwarfs existing in our stellar system is about the same number. This is an independent confirmation of our proposal.

The continuous process of formation of planetary nebulae is the most powerful supplier of gas to interstellar space. A mass of about several tenths of one solar mass per year appears in the Galaxy due to this mechanism. This exceeds by several hundred times the amount of mass ejected by all the novae.

As the luminosity of the planetary nebulae decreases, their optical thickness beyond the Lyman limit becomes very small. All recent methods for the determination of the temperatures of planetary nebulae gave a much underestimated value of T_*. The large abundance of hydrogen in the planetary nebulae is of interest for the theory of white dwarfs.

It is important to establish the nature of stars that create the planetary nebulae. If the development is extrapolated back to the early stages of expansion, the stars must have been objects of high luminosity. The nebular lines in their spectra must then have been fainter than the hydrogen lines and other permitted lines, because the electron concentration was then relatively high. The optical thickness beyond the Lyman limit must have been much greater than unity. Such objects must possess quite bright continua, which originate from two-photon transitions from the 2^2S level of hydrogen. Objects similar to planetary nebulae, but with spectra similar to some Be stars, have been observed by Dr Minkowski. It is clear, however, that unrestricted extrapolation is not safe in this case.

Although the conditions prevailing in the earliest stages of the formation of planetary nebulae are not known, one has the impression that the immediate predecessors of the planetary nebulae may be some type of peculiar red giant star of high luminosity. It is probable that such stars are known, but there is as yet no suspicion that these are the 'proto-planetary nebulae'. These peculiar stars of high luminosity must be a definite, regular, but rather short-lived stage in the evolution of an extremely numerous type of star. It is natural to connect our suggestion with the theory of stellar evolution that has been developed by Schwarzschild and others.

Special observations may prove and extend these considerations. A new determination of the trigonometric parallax of NGC 7293 would be particularly important.

BIBLIOGRAPHY

I. S. Shklovsky, *A.J.U.S.S.R.* **33**, 222, 315 (1956).

DISCUSSION

In the period following the papers on instability among the hot stars of low luminosity, Dr L. Gratton presented a brief account of work in progress on η Carinae by Dr Platseck, Miss Ringuelet, and himself at Córdoba and La Plata. The bright line spectrum of η Carinae is being studied in detail; one interesting observation is that the emission lines of Ti II are quite weak as compared to their intensities in novae of comparable excitation. The Balmer lines are also faint. The hydrogen emission lines consist of fairly wide structures upon which are superimposed narrow emission cores, as well as absorption features. The La Plata workers find that the narrow emission lines originate in the nuclear star, while the bright lines that arise from the nebulosity in the immediate vicinity are very wide, their breadths corresponding to a velocity range of 500 to 1000 km./ sec. The radial velocities vary greatly from one part of the nebula to another. Gratton and his colleagues believe that η Carinae is actually a member of the Carina O-association upon which it is seen superimposed. The corrected distance modulus of this complex is 12·3 magnitudes, with an uncertainty of less than half a magnitude. It follows that η Carinae at its maximum in 1843 had an absolute magnitude almost certainly exceeding -13, and even now is a fairly luminous object of $M = -5$ or -6. The La Plata astronomers believe it to be improbable, therefore, that η Carinae is a nova-like variable or a super-nova, but think rather that the object is to be regarded as similar to S Doradus or Hubble's variables in extragalactic nebulae. They comment on the probable membership of both S Doradus and η Carinae in O-associations, and suspect that this connexion may be of great evolutionary significance.

Dr E. Kharadze stated that he found the report of Walker on the light variations of old novae to be very interesting and important, and described recent work of his own on P Cygni, an object of interest from this point of view. Kharadze believes it is possible that P Cygni, because of its large dimensions and high luminosity, might suffer disturbances of the equilibrium between gravity and light or gas pressure, at least in its outer layers. He and Dr Magalashvili, at the Abastumani Observatory, have been making regular observations of the brightness of P Cygni since 1952, and have found variations exceeding 0·2 or 0·3 magnitude. For example, in July 1952 a decrease of light by 0·2 magnitude was observed, but the brightness recovered in the following 10 days. Similar, or even more striking variations occurred in 1954. In yellow light the light variation was noticeably prompter. Dr Kharadze commented that the magnitudes of P Cygni, and of P Cygni-like stars, deserve regular observation.

Dr E. Schatzman drew attention to the theory he has proposed for the origin of the novae, and which seems to him to be confirmed by the work reported by Walker. The novae must be stars very near a state of instability, and a secular change of structure would lead to an explosion. Schatzman pointed out also that should a white dwarf approach instability, the period of its pulsation would be of the order of 2 sec.

III. INSTABILITY IN THE LUMINOUS STARS
OF EARLY TYPE

13. ON THE ATMOSPHERES OF THE WOLF-RAYET STARS

B. A. VORONTSOV-VELYAMINOV

Sternberg Astronomical Institute, Moscow, U.S.S.R.

In 1945 we showed[1] from a study of six Wolf-Rayet stars that their spectrophotometric temperatures, when allowance is made for interstellar reddening, are still far below the temperatures as determined by the Zanstra method. It was demonstrated that these spectrophotometric temperatures (which are only slightly in excess of those of Ao stars) agree perfectly with N. Kosyrev's[2] theory of extended photospheres. Later Petrie[3] confirmed our observational results, though he did not compare the two sets of results for individual objects. This same result was recently obtained by Andrillat[4].

New plates which we obtained in 1946 in Abastumani lead to temperatures that are in good agreement with our former measurements. Four additional stars were studied with the same general result. Therefore the low spectrophotometric temperatures of the Wolf-Rayet stars are fairly well established.

It was natural to compare the conclusions concerning the extension of the Wolf-Rayet atmospheres obtained from these data with the hypothesis of outflow of matter due to Beals. The latter successfully explained the great width of the Wolf-Rayet spectral bands.

New facts accumulated during the last decade show that this hypothesis meets with some difficulties on account of the very large radii it predicts for the Wolf-Rayet atmospheres. Some of these difficulties were first noticed by O. Wilson[5], then by C. Beals himself[6], and have recently been summarized by M. Johnson[7]. The difficulties enumerated by the latter may be supplemented by the following:

(1) Notwithstanding the supposed intense outflow of atoms from the atmospheres of the Wolf-Rayet stars (except the dwarfs, which are nuclei of planetaries), as a rule these stars are not surrounded by visible nebulosities, and no forbidden lines are present in their spectra. (We would expect the ejected matter to form observable nebulosities after some lapse of time.)

(2) Many dwarf stars, such as the nuclei of planetaries and ex-novae, having very large gravity at their surfaces show the Wolf-Rayet characteristics in their spectra.

(3) The light pressure does not account for the observed motions of gases in the atmospheres.

(4) Appreciable variations of brightness are lacking. This is evidence against the importance of the role of irregular explosive effects in the atmospheres of the stars in question.

(5) A rapid loss of matter would result in a considerable dispersion of masses among the stars of this class, yet a large dispersion is apparently missing.

Therefore, attempts to find another explanation for the phenomena displayed in the Wolf-Rayet spectra are welcomed. The hypothesis of turbulence must, however, be abandoned since for the velocities required the turbulence will be equivalent to ejection.

Whether the physical explanation of the solar flares as discharges in an ionized gas is correct or not, it is possible that if large areas of the surfaces of the Wolf-Rayet stars are spotted by such flares, the scattering of the atomic emission in the flares by fast electrons could account for the large observed width of the Wolf-Rayet bands. If on the relatively quiet solar surface the width of the Hα line in flares amounts to 21 angstroms, there must exist stars in which this broadening is still larger and the flares more frequent and more numerous.

Such a hypothesis might be proven by an attempt to calculate theoretically the relative width and intensities of spectral bands due to different ions and atoms. These ought to be compared with the observed ones, which hitherto were interpreted in terms of stratification of atmospheres, although the latter can actually exist apart from prominent flares.

Estimates of the age of the Wolf-Rayet stars based on the rate of loss of matter will be meaningless in the case of a definite rejection of the expansion hypothesis.

REFERENCES

[1] B. A. Vorontsov-Velyaminov, *A.J. U.S.S.R.* **22**, 93 (1945).
[2] N. A. Kosyrev, *M.N.* **94**, 430 (1934).
[3] R. Petrie, *Publ. Dom. Astrophys. Obs. Victoria*, **7**, 383 (1948).
[4] Y. Andrillat, *Comptes rendus*, **239**, 480 (1954).
[5] O. Wilson, *Ap.J.* **95**, 402 (1942).
[6] C. Beals, *Publ. Dom. Astrophys. Obs. Victoria*, **6**, 95 (1934).
[7] M. Johnson, *Observatory*, **74**, 124 (1954).

14. INSTABILITY IN STARS OF EARLY SPECTRAL TYPE

OTTO STRUVE

Department of Astronomy, University of California, Berkeley, California, U.S.A.

This paper consists of two parts which, at first sight, are quite detached from each other: the pulsating variables of spectral types B to F, and the close spectroscopic binaries of the types of β Lyrae, UX Monocerotis and

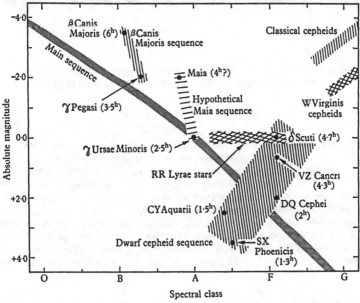

Fig. 1. The sequence of pulsating stars in an absolute magnitude *versus* spectral class diagram.

U Cephei. I shall show that there exists a connexion between these groups which, though still rather nebulous (in both senses of this word), promises to yield interesting results.

Fig. 1 shows the various sequences of pulsating stars with which we have been concerned in recent years at Berkeley. The sequence of the β Canis Majoris stars has been explored in great detail. There is a pronounced period-luminosity relation and also a period-spectrum relation. We are at present working on two stars of spectral type B3 which have even shorter periods than γ Pegasi, but the final conclusions are not yet available;

one, 53 Piscium, may vary in light with $P \approx 2\frac{1}{2}$ hours, according to A. D. Williams, and in velocity with $K \approx 10$ km./sec., according to R. T. Mathews.

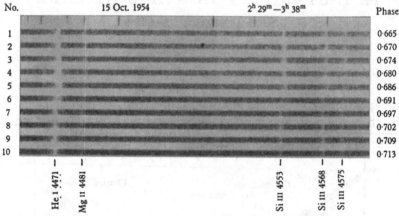

Fig. 2a. Spectra of BW Vulpeculae in the $\lambda\lambda$ 4460–4585 region, obtained by A. J. Deutsch. The phases in the margin correspond to the horizontal scale of Fig. 3.

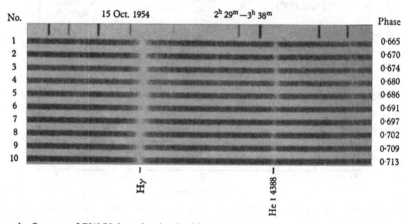

Fig. 2b. Spectra of BW Vulpeculae in the $\lambda\lambda$ 4300–4425 region, obtained by A. J. Deutsch. The phases in the margin correspond to the horizontal scale of Fig. 3.

The most interesting result pertains to the discontinuity of the velocity curves of BW Vulpeculae, σ Scorpii, 12 (DD) Lacertae, and other members of this sequence. It is almost certainly a universal phenomenon. Fig. 2 shows several spectra of BW Vulpeculae, obtained by A. J. Deutsch with the 200-inch telescope. They show the duplicities of the absorption lines much better than my own spectrograms of this star which were obtained in September 1954 at Mount Wilson. Notice that all lines split into

94

components, but that there is a slight delay in Hγ, as compared to He I λ 4472 and the lines of Si III. On the sixth (from the top) exposure the violet component of He I and Si III is already the stronger, while on the same exposure the red component of Hγ is still the stronger of the two. I have referred to this delay as the 'Van Hoof effect', since it was discovered by Dr A. Van Hoof. Fig. 3 shows the velocity curve of

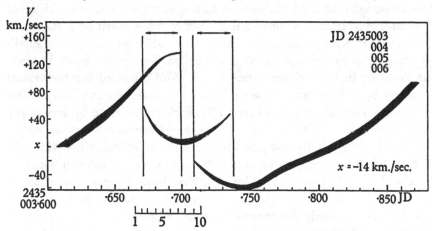

Fig. 3. The velocity curve of BW Vulpeculae. The phases of the spectrograms reproduced in Figs. 2a and 2b are indicated by the scale at the bottom. The origin of the velocity ordinates, *x*, corresponds to − 14 km./sec.

BW Vulpeculae, of $P=4^{\rm h}\ 49^{\rm m}$. The character of the discontinuity is not the same as in RR Lyrae and W Virginis, but its theoretical implication is probably similar. Notice that the β Canis Majoris stars are members of population I.

It is reasonable to say that what we observe in these stars are strictly periodic puffs of gas whose velocities are decelerated by gravity, which causes the gases to fall back upon the surfaces of the stars. We have, then, on a very small scale, something in the nature of nova explosions. For further details, I refer to the work of G. J. Odgers at Victoria.

We are, at the present time, engaged in the study of several stars of the δ Scuti sequence, especially HD 199908. These stars resemble in many ways the β Canis Majoris variables, but they also resemble the classical RR Lyrae or cluster-type variables. In all three sequences we frequently find pronounced beat phenomena. Delta Scuti shows a broadening of its absorption lines on the descending branch of the velocity curve. This may mean that its velocity curve would also be discontinuous if we could resolve the components, but intrinsic broadening by turbulence or rotation prevents us from doing this.

95

The sequence of Maia is quite uncertain and we shall not be able to explore it in the near future.

I now turn to the binaries. The connecting link between them and the pulsating variables is provided by three recent investigations made by my colleagues at Berkeley. The first is a study of UX Monocerotis by C. R. Lynds who has shown that the smaller, A-type, component of this binary of 6-day period is an intrinsic variable having some of the characteristics of the cluster-type stars. The second is a study of AE Aquarii by J. A. Crawford and R. P. Kraft, which attributes the nova-like outbursts of its small, hot component to the capture of gas expelled through the inner Lagrangian point by its large K-type companion. And the third is a theoretical discussion by S. S. Huang of the evolutionary consequences of loss of mass in close binaries. He suggests that while single stars evolve by way of the cluster-type sequence, the components of close binaries by-pass this stage: they evolve more rapidly and pass through a region of instability which, in the H–R diagram, occupies an area of triangular shape between the post-novae and the RR Lyrae variables. The details of this work will soon be published. It leads to a logical interpretation of the duplicity of Nova Herculis 1934, recently discovered by M. F. Walker.

Because of the importance of this problem I have recently started, in collaboration with J. Sahade, a new study of the spectrum of β Lyrae. Figs. 4a to 4f show the cyclic variations of the spectrum in its 12·9-day period for two regions of the spectrum (near the K line of Ca II, near Hγ, and near He I λ 4472). One set of plates was lined up for the lines of the B9 star, while the other was adjusted by means of the interstellar lines H and K. The phases are shown as decimal fractions of the period, and the numbers which precede the decimal point identify the cycle. The spectrum at the top, phase 2·6451, was obtained on 9 May 1955, at $8^h 27^m$ U.T.

This complexity of the variations is very great, and my purpose in presenting this material is to solicit help in interpreting it. In a general way, my earlier interpretation was probably correct, but there are many interesting features which were not seen on the relatively inferior Yerkes plates.

I have time to mention only a few of the more remarkable results of our present work:

(1) The satellite absorption lines are of special interest just before mid-eclipse and immediately after it. The former are narrower than the latter, but the high-dispersion plates show that the former are also broadened by Doppler motions. Their edges correspond, in the mean, to velocities of $+140$ to $+200$ km./sec., but these velocities increase as we pass from phase 0·95 to phase 0·98, approximately, and in all probability the range also increases.

(2) It was at first disturbing that the spectrum of these redward satellites appeared to resemble rather closely that of the B9 star. There was no reason to believe that this should be the case. However, we can now definitely state that the spectral class as defined in the conventional manner by these satellites is considerably later than B9. It is more nearly A2: the Ti II lines are definitely enhanced (notice especially Ti II λ 4501 which shows a well-marked satellite but no corresponding B9 component); Ca II is also enhanced but not enough to make the spectral type much later than A2. The stream which gives rise to these satellites probably has a higher degree of ionization and excitation than the invisible atmosphere of the companion star. We can infer only that the spectral type of the latter is later than A2.

(3) There is some suggestion that immediately after the disappearance of the red absorption satellites they are replaced by weak emission components. This is best seen in Si II, Mg II and perhaps Ca II. After mid-eclipse the main mass of this stream is hidden by the companion star. Only a fringe remains visible, but this is not seen projected upon the disc of the B9 star.

(4) The violet absorption satellites are strong and very broad. They appear suddenly right after mid-eclipse and range in velocity between -80 and -360 km./sec., remaining about the same between phases 0·04 and 0·05. But at phase 0·09 and until phase 0·12 they are replaced by fairly strong emission components. Evidently this stream remains projected against the B9 disc until phase 0·09, after which it is seen partly against the space between the two stars, and partly against the disc of the invisible F companion star.

(5) This stream is strong in H and He I, but it also shows lines of N II, Ca II and Fe III (λ 4419·6). There seems to be little or no Fe II, Ti II, and strangely no O II. There is probably no He II and no Si III. There is little or no dilution effect: the satellites of He I λ 3888 and λ 3965 are not enhanced, as are the corresponding 'fixed' B5 lines. No doubt this stream is fairly close to the surface of the B9 star, but its ionization is considerably greater, probably because of reduced density and increased temperature.

(6) We conclude that these streams are produced by the instability of *both* component stars near the inner Lagrangian point, a phenomenon which G. P. Kuiper has designated as instability of class A. This agrees with the theoretical results by V. A. Krat, A. N. Dadaev and A. Kranjc. It also permits us to connect the phenomena in β Lyrae with those in U Cephei, U Sagittae, SX Cassiopeiae, UX Monocerotis and many other

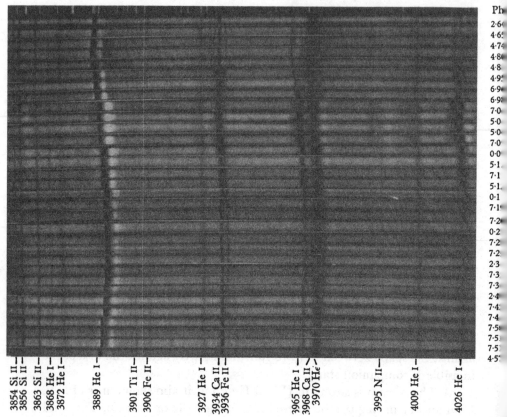

Fig. 4*a*. Spectra of β Lyrae in the region $\lambda\lambda$ 3850–4030. The individual spectra have been shifted so that th features of the B9 component are aligned throughout the series. The phases are in decimal fractions of the perio and the numbers which precede the decimal point identify the cycle.

98

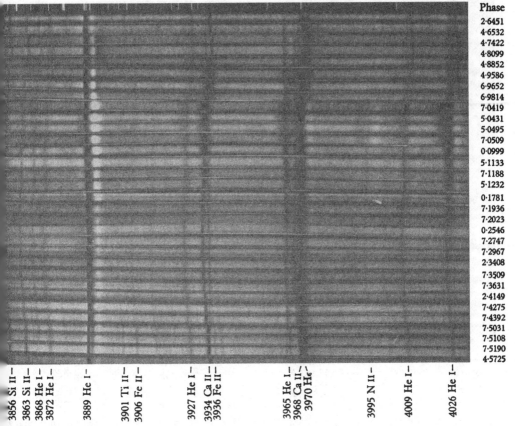

	Phase
	2·6451
	4·6532
	4·7422
	4·8099
	4·8852
	4·9586
	6·9652
	6·9814
	7·0419
	5·0431
	5·0495
	7·0509
	0·0999
	5·1133
	7·1188
	5·1232
	0·1781
	7·1936
	7·2023
	0·2546
	7·2747
	7·2967
	2·3408
	7·3509
	7·3631
	2·4149
	7·4275
	7·4392
	7·5031
	7·5108
	7·5190
	4·5725

3856 Si II – 3863 Si II – 3868 He I – 3872 He I – 3889 He I – 3901 Ti II – 3906 Fe II – 3927 He I – 3934 Ca II – 3936 Fe II – 3965 He I – 3968 Ca II – 3970 He – 3995 N II – 4009 He I – 4026 He I –

ɼ. 4*b*. Spectra of *β* Lyrae in the region λλ 3850–4030. The spectrograms are the same as those used in Fig. 4*a*, except that the alignment has now been made with the aid of the interstellar H and K lines.

7-2

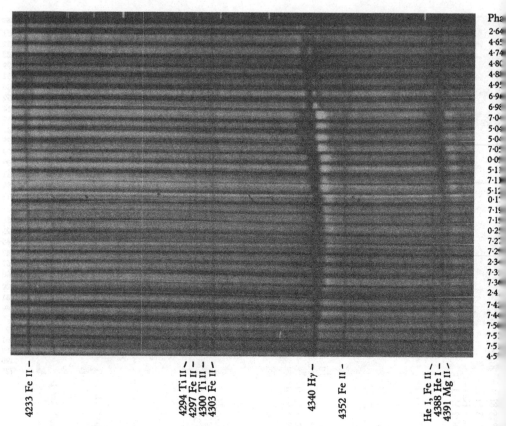

Fig. 4c. Spectra of β Lyrae in the region $\lambda\lambda$ 4230–4400. The individual spectra have been shifted so that the features of the B9 component are aligned throughout the series. The phases are in decimal fractions of the period and the numbers which precede the decimal point identify the cycle.

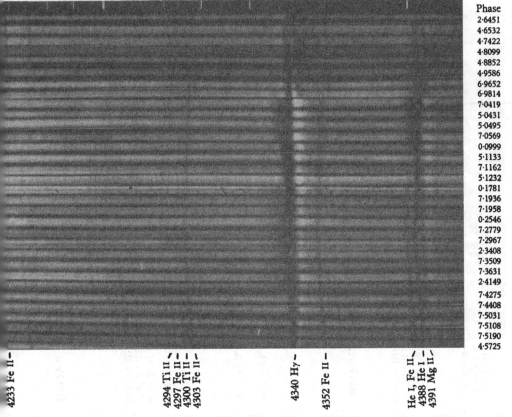

Phase
2·6451
4·6532
4·7422
4·8099
4·8852
4·9586
6·9652
6·9814
7·0419
5·0431
5·0495
7·0569
0·0999
5·1133
7·1162
5·1232
0·1781
7·1936
7·1958
0·2546
7·2779
7·2967
2·3408
7·3509
7·3631
2·4149
7·4275
7·4408
7·5031
7·5108
7·5190
4·5725

4233 Fe II —

4294 Ti II
4297 Fe II
4300 Ti II
4303 Fe II

4340 Hγ —

4352 Fe II —

He I, Fe II
4388 He I
4391 Mg II

4d. Spectra of β Lyrae in the region λλ 4230–4400. The spectrograms are the same as those used in Fig. 4c, except that the alignment has now been made with the aid of the interstellar H and K lines.

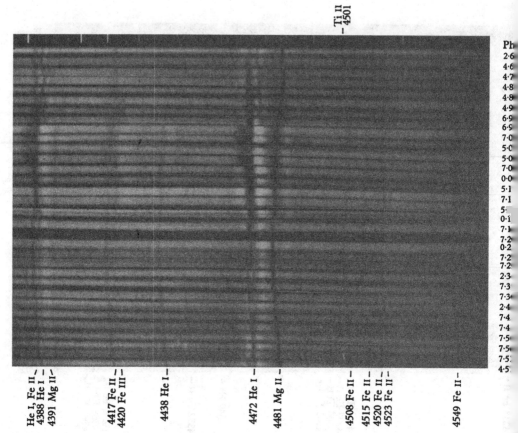

Fig. 4e. Spectra of β Lyrae in the region λλ 4380–4560. The individual spectra have been shifted so that the features of the B9 component are aligned throughout the series. The phases are in decimal fractions of the period and the numbers which precede the decimal point identify the cycle.

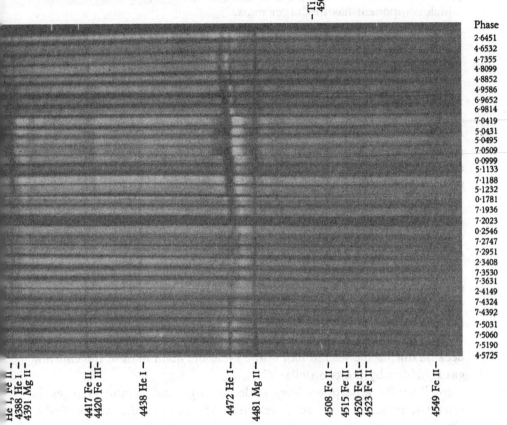

Phase
2·6451
4·6532
4·7355
4·8099
4·8852
4·9586
6·9652
6·9814
7·0419
5·0431
5·0495
7·0509
0·0999
5·1133
7·1188
5·1232
0·1781
7·1936
7·2023
0·2546
7·2747
7·2951
2·3408
7·3530
7·3631
2·4149
7·4324
7·4392
7·5031
7·5060
7·5190
4·5725

4f. Spectra of β Lyrae in the region λλ 4380–4560. The spectrograms are the same as those used in Fig. 4e, except that the alignment has now been made with the aid of the interstellar H and K lines.

103

close binaries. Evidently this type of instability occurs always when one component star (or both) is spilling over its critical zero-velocity equipotential surface containing the Lagrangian point L_1. It does not matter which component has the larger mass.

(7) In our interpretation of the emission and absorption lines of β Lyrae we must remember that they are not necessarily, and not even usually, produced in the same layers of gas. The importance of this point was brought to my attention twenty years ago by the late B. P. Gerasimovič. Let us consider first the broad emission feature of He I λ 4472 (the other He I lines, excepting those whose lower levels are metastable, and the H lines behave in a similar manner). This feature is produced not only by the condensation of gas between the Lagrangian point L_1 and the following side of the F star, but also by a large cloud of gas which envelops the entire system. Nevertheless, orbital motion is evident from the curvature of the displacements when the spectra are aligned according to the stationary lines. This curvature is conspicuous between phases 0·0 and 0·5, and it is opposite in sense to that shown by the B9 lines: the main mass of the emitting gas moves in phase with the F star and out of phase with the B9 star. The value of K is roughly 100 km./sec., that of the B9 star is 180 km./sec., and that of the F star is 270 km./sec., if Kuiper's mass ratio $m_{B9}/m_F = 1\cdot5$ is correct. The orbital motion of the main mass of the emitting atoms is in agreement with A. A. Belopolsky's early work of 1892.

(8) This orbital motion is not detectable between phases 0·5 and 0·0 because during the second half of the cycle the main mass of the emitting gas is hidden by the two stellar discs.

(9) The B5 absorption lines undergo large and erratic changes in intensity, structure and radial velocity. My earlier work at the Yerkes Observatory indicated that these lines—always strongly displaced shortward—are displaced less before mid-eclipse than after it. The new observations do not disprove this. J. G. Baker's measurements of He I λ 3888 and λ 3965 also show this effect. Since these lines are produced by the gas which is between us and the B9 disc, they do not indicate the motion of the streams except very close to mid-eclipse. They appear then to be *in phase* with the B9 star, despite their large negative displacements. Evidently the large shell enveloping the system expands and revolves around the entire system; the shell is concentric, not with the B9 star, but with the centre of gravity of the system. I believe that this, too, is compatible with Krat's theory. The observed component of the orbital velocity of the shell is small compared to the orbital motions of the stars and to that of the condensation discussed previously.

(10) One of the most conspicuous features of the emission in He I λ 4472 is the sudden increase in intensity at mid-eclipse. This is not nearly so conspicuous in He I λ 3888 and λ 3965. Evidently the uncovering of the emitting condensation—rich in He I λ 4472 and poor in He I $\lambda\lambda$ 3888, 3965—is responsible for this effect. This is in harmony with the results of paragraph (9).

(11) The lines of He I $\lambda\lambda$ 3888, 3965 are produced mainly in the large shell, but this is more the case for the former line than for the latter. The line He I λ 3888 blocks out almost completely the line Hζ. The intensity of the emission line He I λ 3888 undergoes considerable variation, but part of it is caused by the increased exposure times during the eclipses. After this effect is removed there remains a *small* increase in intensity at mid-eclipse which is, no doubt, the same phenomenon as in He I λ 4472. There is also an asymmetry: the intensity is somewhat lower at phase 0·88 than at phases 0·25 to 0·40. This is qualitatively consistent with the observations of K. Saidov. But, contrary to his results, we find that the emission line is never absent. We cannot at this time venture any comments regarding the mechanism proposed by Saidov to account for the variation. It is not clear to us whether his observations were corrected for the exposure-time effect.

(12) Those B9 lines of Si II, Ca II, Mg II, and Fe II which are not complicated by emission behave in a fairly normal manner. But they are considerably strengthened at secondary minimum, which may be due to the absence of the continuous light of the F star. Measurements of this strengthening should give a determination of L_{B9}/L_F, perhaps even as a function of λ.

(13) During the principal eclipse the B9 lines of N II and He I, which are probably not noticeably complicated by emission, are slightly reduced in intensity, without being strengthened to an equal extent at secondary eclipse. We shall not now attempt to explain the distinction between (12) and (13).

(14) The fainter B9 lines become broad and hazy at certain phases during principal eclipse. One thinks immediately of Rossiter's rotational disturbance. But this effect is small, about $\pm$ 12 km./sec. in radial velocity. Moreover, the equatorial rotational velocity of the B9 star is 40 km./sec., as determined from the profiles observed outside of eclipse. The broadening at mid-eclipse is not the same in different cycles, but it is always more pronounced after mid-eclipse than before, and it seems to produce lines which are about twice as broad as is consistent with the rotation of the B9 star. There is some shading toward the violet, after mid-eclipse, especially in Si II $\lambda\lambda$ 4128, 4132. This is almost certainly an effect of electron scattering

in the highly ionized stream of gas which flows from the B9 star toward the observer. The temperature and electron density are of the right order of magnitude to produce the observed broadening, as S. S. Huang has suggested; and the macroscopic motion of the stream in front of a considerable portion of the B9 disc would, according to Mrs E. Böhm-Vitense, produce a small asymmetry in the line—qualitatively in accordance with the observations.

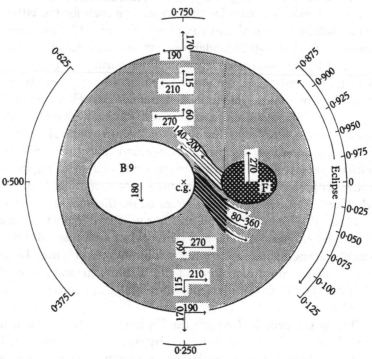

Fig. 5. A model of β Lyrae.

(15) There are similar, but less pronounced irregular changes in the profiles of the same lines (especially Fe II λλ 4508, 4515, 4520, 4523, etc.) outside of eclipse. We do not yet know whether these changes affect the radial velocities. They could also be due to electron scattering (or to changes in turbulence) in the large shell.

(16) The puzzle of the rotational velocity of the B9 star remains unsolved. Since $a_1 \sin i = 33 \times 10^6$ km., and $R_1 = 0.5 \, (a_1 + a_2)$, it is impossible to reconcile the observed velocity with the value of $2\pi R_1 / P_{\text{orbit}}$, for which Kopal's elements give 230 km./sec. This difficulty remains even when the mass ratio is quite different from 1·5, as given by Kuiper. If we abandon the mass-luminosity relation for the F star and assume that $m_1 / m_2 < 1$, the

rotational velocity can be reduced to about 100 km./sec. But it is still greatly in excess of the observed value. Perhaps we must assume that $P_{\text{rot.}} > P_{\text{orbit}}$. This would be unusual in an eclipsing binary; but, on the other hand, it is known that single super-giants rotate slowly. It may, of course, be that the outer strata rotate more slowly than the inner.

(17) Still another puzzle is that of the luminosity of the B9 star. The spectroscopic criteria, such as the number of H absorption lines and their Stark broadening, suggest $M_{v,1} = -4\cdot5$; the mass-luminosity relation for the B9 star would then give $m_2 = 13$ m⊙. The mass function

$$m_2 \sin^3 i/(1+\alpha)^2 = 8\cdot3 \text{ m⊙}$$

would give $m_2 = 21$ m⊙ and $\alpha = m_1/m_2 = 0\cdot6$. This does not help with the rotational discrepancy of the B9 star, and we do not believe that the result is correct. But it is also difficult to accept Kuiper's compromise solution $M_{v,1} = -7\cdot6$ and $\alpha = 1\cdot5$, since his other criteria suggest a lower luminosity.

(18) Fig. 5 summarizes the results obtained from the observations. The spectrograms of β Lyrae were obtained by O. Struve and J. Sahade as guest investigators at the Mount Wilson Observatory. All spectrograms, 190 in number, were made with the coudé spectrograph of the 100-inch telescope, using Eastman Process emulsion. The dispersion was 10 Å/mm.

DISCUSSION

In the discussion of the papers of Section III, Dr L. Gratton stated that in the case of β Lyrae and other close binaries, one may expect that the radii of the components are considerably larger than those of single stars of the same mass. This would cause the central temperatures to be lower and, in turn, the luminosities to be smaller. It might even turn out, according to Gratton, that in close binaries the generation of energy might depend on the p–p reaction rather than on the C–N cycle, notwithstanding the large masses of such stars.

Dr W. P. Bidelman suggested in connexion with the question of the absolute magnitude of β Lyrae, that a reliable value of this quantity could almost certainly be obtained from the spectroscopic absolute magnitude of the visual companion. He believed that the visual absolute magnitude of β Lyrae derived in this way was of the order of -4 to -5, but stated that a more reliable value could be determined.

Dr A. Van Hoof made reference to Dr Struve's comments on the spectral peculiarities of the β Canis Majoris stars, and described his own discovery, in β CMa itself, of the periodic appearance and disappearance of two lines at $\lambda\lambda$ 4818·3 and 4845·6. At about the time of maximum amplitude of the radial velocity variation, these two lines are of about the same strength as lines of the nearby Si III triplet. But at about the time of minimum velocity amplitude, these new lines are completely invisible. In other words, the pair is present when the short-period variations (P_1 and P_2) are in phase, but vanishes when these variations are out of phase. The λ 4818 line has a normal profile for its intensity, but λ 4845 is too wide for its strength. The origin of these lines is obscure, according to Van Hoof. It would be interesting to find if other stars of this type show the same features.

Dr Martin Johnson posed a question whose answer he felt might tie together the papers of Dr Mustel, Dr Struve, and Dr Walker. The contributions of the last two authors discussed the behaviour of binary stars and the circumstellar material in their vicinities, and Dr Mustel made the interesting suggestion that novae possess strong magnetic fields. Dr Johnson inquired what changes in the ordinary mode of flow of the circumstellar gas streams would follow if one or both stars were to have strong magnetic fields, and if these fields were parallel, or antiparallel, or oriented obliquely. An answer may be difficult to make, but Dr Johnson believed it might bear significantly upon the possible causes of outbursts, since electro-magnetic effects are able to amplify small disturbances.

IV. INSTABILITY IN THE REGULAR VARIABLES OF LATER TYPE

15. VARIABLE STARS AND PROBLEMS OF STELLAR FORMATION

B. V. KUKARKIN

Sternberg Astronomical Institute, Moscow, U.S.S.R.

The possibility of discovering phenomenologically similar objects either located in different stellar systems, or in totally different (according to their origin and age) parts of some complex stellar system (as, for instance, our Galaxy) is of extreme importance. The detection of such objects permits us to confirm that, in spite of different initial conditions and evolutionary paths, stars of quite different origin pass in the course of their evolution through the same stages.

Variable stars are such phenomenologically similar objects. They might be called 'marked stars', since it is rather easy to recognize and distinguish them from other stars because of the specific properties of their variability. Mira Ceti type stars are particularly interesting and promising. The large variations of their brightness make their discovery and study extremely simple. Of prime importance is the fact that Mira Ceti type stars are found in most different parts of the Galaxy (in the flat, spherical, and intermediate components) and in isolated systems of various types (as, for example, in globular clusters and the Magellanic Clouds). The stars contained in various galaxies, or in different parts of our Galaxy, are doubtlessly of quite different origin.

The preliminary analysis of the morphological properties of Mira Ceti type stars in different systems and in different components of our Galaxy led to the decisive conclusion that these properties are extremely diverse [1]. Thus, in the Magellanic Clouds the periods of all the Mira Ceti stars exceed 540 days, whereas in globular clusters the periods are less than 310 days. The light curves of the Mira Ceti stars belonging to the flat component of our Galaxy are asymmetrical and have a clearly expressed hump on the ascending branch, while the light curves of stars in the spherical component are symmetrical and only some of them show a slight wave on the ascending branch. Many more examples of this kind might be given.

Similar stages of stellar evolution are observed in different stellar systems and in various parts of complex stellar systems. For every system (or for parts of a complex stellar system) the Mira Ceti stars (as well as variable stars of other types) show slight morphological differences.

The Mira Ceti stars are bolometrically among the brightest stars. The absolute bolometric magnitude of the Mira Ceti stars exceeds − 3, reaching even − 7 or − 8 for some members. This suggests that the Mira Ceti stars represent quite massive and, therefore, quite young stars.

The Mira Ceti stars show well-known and extensively studied spectroscopic anomalies. Among these anomalies the emission spectrum is the most striking. The intensity of the photospheric radiation which determines the absorption spectrum of the Mira Ceti stars (a temperature of the order of 2500° K. is representative) can by no means produce the emission spectrum. The emission spectrum can be caused, as was convincingly shown by G. A. Shajn [2], only by strong short-wave radiation, the intensity of which must exceed the short-wave radiation of the photosphere by many orders of magnitude. Thus, we meet here with short-wave continuous emission, similar to the continuous emission of other, obviously young stars. The cosmogonical significance of the properties of these latter objects has been discussed by V. A. Ambartsumian [3].

P. Merrill [4] has recently discovered in the spectra of some Mira Ceti variables intense lines of technetium. This discovery suggests that Tc originates directly in the outer layers of the Mira Ceti stars and that, consequently, element formation is going on in these layers. Either in the Mira Ceti stars there are conditions particularly favourable for element formation, or the Mira Ceti stars themselves are very young and still in 'the making'. The latter supposition seems to be the most probable.

The apparent distribution of Mira Ceti stars on the celestial sphere is characterized by the existence of real groups and clusters. These visible groups are the reflexion of the actual spatial clustering. The peculiar tendency to form spatial nests is a typical feature of Mira Ceti stars located in the galactic plane and of those located in high galactic latitudes. Taking into account the velocity dispersion of the Mira Ceti stars, it may be shown that the observed groups cannot be of a long duration. This suggests, in its turn, the youth of the Mira Ceti stars.

The above facts testify that the process of star formation is going on not only in the spiral arms of galaxies, but also in stellar systems that many authors are inclined to consider 'quite old' and deprived of their powers of star formation.

Numerous proofs have been obtained during recent years which show that in different stellar systems and in various parts of complex stellar systems, like our Galaxy and M31, variables of the same type possess their own specific morphological properties. Special studies have been devoted to that problem [5, 6, 7, 8].*

* See also Prof. Kukarkin's communication to the I.A.U. Symposium on the Large-Scale Structure of Galaxies, held in Dublin in September, 1955.

These facts suggest that in different parts of the Universe matter passes through phenomenologically similar, but not quite the same, evolutionary stages. If variable stars show slight differences, but are generally quite similar, this may be interpreted as a difference of initial conditions that existed in the stellar system, or in some parts of a complex stellar system. Furthermore, the existence of variable stars of the same type in extremely different stellar systems allows one to believe that the evolution of these systems passes through similar stages: i.e. that their ages are equal.

Variable stars of certain types are apparently young. Thus, the numerous RW Aurigae, or T Tauri, stars form unstable associations and are connected with dark and comet-shaped nebulae [3, 10, 13]. It is evident that in this case we see young stars of recent origin. Semi-regular and irregular super-giant variables also form associations [9, 11].

Important arguments have been put forward in favour of the youth of carbon stars [12] as well. There are reasons to believe that the Mira Ceti stars are also of recent origin [1].

There are no reasons to believe that any variable stars of other types are particularly young. Thus, cepheids do not form associations and are not connected with them. The long-period cepheids are more or less uniformly distributed in the galactic plane, in the nearest spiral systems, and in the Magellanic Clouds. There are no reasons to think that novae, nova-like variables, and red semi-regular variables (not super-giants) are particularly youthful objects.

The existence of both extremely young and older variables permits one to compare their presence with the peculiarities of some stellar systems and their parts. Evidently young stars are met both in spiral arms of the Galaxy and in its spherical component, both in the spiral stellar systems and also in globular clusters and elliptical stellar systems. This permits one to state that the process of star formation is going on not only in the flat, but also in the spherical component of our Galaxy, and in the spiral and the elliptical galaxies.

At the same time in some stellar systems young stars are found, whereas they do not occur in other systems of similar types. It may be concluded that the process of formation of the stellar systems themselves is still going on and that both young and old stars are met in them. If the presence of variables of definite types is actually an extremely certain criterion of the youth of a given stellar system and of the process of stellar formation going on in it, the necessity for a study of the peculiarities of the population in many such systems becomes quite apparent.

The colour-magnitude diagram permits one to characterize the population of a stellar system rather fully and objectively. It is well known that this diagram has a quite different character for the flat and spherical components of our Galaxy. But we also know that the process of stellar formation is going on in both the first and the second components. The above-mentioned sharp difference is, consequently, caused by the dissimilarity of the initial conditions in the formation of stars in these two components.

When studying the peculiarities of the colour-magnitude diagrams for various stellar systems in the same component (for example, for different globular or open clusters), we discover, along with a general likeness, an extreme dissimilarity in their details (as, for example, in passing from one globular cluster to another). Such a dissimilarity is usually connected with the presence or absence of variable stars of different types. For example, the colour-magnitude diagrams for globular clusters that are rich in variables of the RR Lyrae type, differ in detail from such diagrams for clusters containing few RR Lyrae stars. This difference is due mainly to the age and not to the initial conditions. The youth of a star cluster in a galaxy is manifested in the peculiarities of the colour-magnitude diagram, and also in the presence of variable stars of definite types.

From the remarkable study of variable stars in the galactic nucleus by W. Baade and S. Gaposchkin [14] it may be concluded that the lifetime of the RR Lyrae stars cannot be long. Their morphological properties are sharply expressed and entirely different from RR Lyrae stars in the solar neighbourhood. At the same time, owing to the extremely large eccentricities of their orbits they should have attained the solar neighbourhood. Their lifetime is probably much less than the period of one galactic rotation. Thus, this is additional proof that within the nucleus of the Galaxy and on its periphery, stars of different origin enter into a similar stage of evolution.

Further investigations of variable stars contained in different stellar systems are required. It is necessary to compare the presence of variables of different types with the peculiarities of the populations in these stellar systems. We are apparently approaching the solution of the main and most difficult problem of stellar cosmogony—that of a separation of the influence of the initial conditions from the properties acquired in the process of evolution. If we remember the simplicity of the methods of investigation and of the precise classification of variable stars, the course that we should follow and along which our efforts should be directed in the near future becomes quite clear and evident.

REFERENCES

[1] B. V. Kukarkin, *A.J. U.S.S.R.* **31**, 489 (1954).

[2] G. A. Shajn, *Comptes rendus de l'Acad. d. Sci. de l'U.R.S.S., série phys.* **9**, 161 (1945).

[3] V. A. Ambartsumian, *Comm. Burakan Obs.* No. 13 (1954).

[4] P. W. Merrill, *Ap.J.* **116**, 21 (1952).

[5] B. V. Kukarkin, *The Study of the Structure and Evolution of Stellar Systems*, Moscow, 1949.

[6] B. V. Kukarkin, *Erforschung d. Struktur und Entwicklung der Sternsysteme*, Berlin, 1954.

[7] B. V. Kukarkin and P. G. Kulikovsky, *Variable Stars*, **8**, 1 (1951).

[8] A. H. Joy, *Ap.J.* **110**, 105 (1949).

[9] B. V. Kukarkin, *Progress of Astronomical Sciences*, **4**, 183 (1948).

[10] P. N. Kholopov, *Problems of Cosmogony*, v. 1, p. 195, Moscow, 1952.

[11] V. A. Ambartsumian, *Comptes rendus de l'Acad. d. Sci. de l'Arménie*, **16**, No. 3 (1953).

[12] J. J. Ikaunieks, *Variable Stars*, **8**, 393 (1952).

[13] P. P. Parenago, *Publ. Sternberg Astronomical Inst.*, **25** (1954).

[14] S. Gaposchkin, *Variable Stars*, **10**, No. 6 (1956).

16. THE STARS WITH DOUBLE ENVELOPES

V. P. TSESEVICH

Astronomical Observatory, Odessa, U.S.S.R.

In modern astronomy we acquire fundamental knowledge concerning the dimensions and the constitution of the exterior layers of giant stars by the study of eclipsing systems. The red giants which form part of the systems of VV Cephei and ζ Aurigae have been studied thoroughly from the changes that the spectra of these systems undergo when the light of the bright eclipsed star passes through the envelope of the rarefied giant. At the present time it is clear that stars having extended envelopes are not rare. Thus, for example, F. I. Loukatskaya [1] (at the Principal Astronomical Observatory of the Academy of Sciences of the Ukrainian S.S.R.) has studied the eclipsing star AW Pegasi in different parts of the spectrum; she has shown that this star possesses a semi-transparent envelope that produces a considerable part of the eclipse. It turns out besides that certain spectral lines, observed in the spectrum of the bright star, come from the semi-transparent envelope of the less luminous companion.

Shulberg [2], of the University of Odessa, has worked out according to the Kosyrev-Chandrasekhar theory of the structure of extended atmospheres a general method for the determination of the elements of an eclipsing binary system that possesses envelopes; he has shown that there takes place in the envelope of the star V444 Cygni a strong absorption of the light of the eclipsing star, which confirms the existence of semi-transparent stellar envelopes.

The more recent work of the author in collaboration with I. G. Jdanova [3] of the Principal Astronomical Observatory of the Academy of Sciences of the Ukrainian S.S.R. includes an explanation of a new method for discovering extended envelopes in giant stars, which might be applied to the study of physical variable stars. The author is of the opinion that researches pursued in that way will yield important results. As is known, the variables of the RV Tauri type are quasi-periodic over more or less long intervals. They maintain for a long time a rigorous periodicity which then suddenly interrupts itself. After a certain lapse of time, in the course of which the light varies irregularly, the regular variations re-establish themselves with a period approximating the initial period.

The light curve also undergoes changes. It occasionally recalls the light curves of the stars of the β Lyrae type, and sometimes those of the typical cepheids.

In addition, in a great number of stars of the RV Tauri type there have been observed long-period variations in the mean brightness. The study of this phenomenon has established that these slow variations are not an exception, but rather are the general rule.

The following are some characteristic values for stars of the RV Tauri type:

Star	P_1 (days)	P_2 (days)	A_I (mag.)	A_{II} (km./sec.)
DF Cygni	49·8	780·2	—	—
R Sagittae	70·8	1112	0·55	23
RV Tauri	78·7	1224	1·70	19
U Monocerotis	92·3	2320	0·30	40

P_1: Period of the rapid variation.
P_2: Period of the slow variation.
A_I: Amplitude of the slow variation of brightness.
A_{II}: Amplitude of the slow variation of the radial velocity.

The comparison of the light curves with those of colour index and radial velocity, deduced from the observations of Joy and of Sanford, have led us to the following conclusions:

First of all, the rapid variations of the light and of the spectral characteristics reveal the following properties, common to all the stars examined.

(1) A careful colorimetric study of AC Herculis in three spectral regions shows that its colour index undergoes great variations. The maximum variations are produced during minimum light. The variation of the colour index between the limits of $+0·1$ and $+1·7$ magnitudes does not correspond to a change of spectral class from F1 to K4. Considerable deviations from thermodynamic equilibrium are evident.

(2) The curve of variation of the colour index is similar in form to the light curve, if one takes for its maximum the minimum colour index, corresponding to the maximum of the colour temperature. At the same time, the curve of colour index precedes that of light by 0·12 period.

(3) The curve of variation of the radial velocities constructed from the metallic absorption lines is in phase with the light curve, as in the cepheids (if the negative values of the velocities are plotted in the positive sense on the ordinate axis).

(4) The curve of variation of the radial velocities constructed from the absorption lines of hydrogen is in phase with the curve of variation of the colour index, preceding the light curve and that of the metallic-line radial velocities by 0·12 period.

(5) At the moment corresponding to the minimum value of the colour index, many of the stars studied showed emission lines of hydrogen.

All this indicates that the rapid variations of light burst forth in different layers of the extended photosphere of the star. The perturbation that leads a new cycle of variations begins at the base of the photosphere. The advance in phase of the curves of colour index and of hydrogen-line radial velocities, which has been discussed, speaks in favour of this. It is at this time that a considerable gradient forms in the distribution of the velocities in the interior of the star which leads to the appearance of the emission lines of hydrogen. Afterwards the perturbation passes into the higher layers of the photosphere. The wave which it propagates rises from the interior of the star and moves out to the periphery.

Following this, we analysed the slow variations of brightness and radial velocity. As is known we are fully informed on the slow variations of the radial velocity only in the case of U Monocerotis, thanks to the researches of R. Sanford. We have succeeded in discovering, by a harmonic analysis of the observations of A. Joy, the slow variation of the radial velocity in R Sagittae and RV Tauri, and to suspect the existence of similar variations in DF Cygni and SS Geminorum. We have studied also the slow variations of brightness that are synchronized with the variations of the radial velocity. As a result, we have obtained some ideas concerning the slow variations of stars of the RV Tauri type. The results can be summarized as follows:

(1) The curves of brightness and of radial velocity are in phase, if the negative values of the velocities are taken in the positive sense on the ordinate axis. Therefore, one cannot suppose that the slow variations of the velocity are caused by orbital motion; it follows that the slow variations of the radial velocities and of light are due to pulsation.

(2) The curves of the radial velocities yield, after their integration, the curves of variation of the radius of the reversing layer. It appears that in RV Tauri the radius varies by 380 million kilometres, in R Sagittae by 374 million, and in U Monocerotis by 1250 million kilometres.

(3) Such large variations of the radius of the reversing layer are possible only in case the radius attains a value of several thousand million kilometres. It follows that the stars of the RV Tauri type are the largest of all those which we know.

The structure of a star of the RV Tauri type can be represented as follows: in the interior there exists a nucleus that possesses a photosphere extending to a radius nearly a hundred times greater than that of the Sun. This nucleus is surrounded by an enormous upper layer, which is an extended reversing layer having an outer radius of several thousand million kilometres.

A mechanical pulsation probably could not be the cause of the variations described above; it is possible that these variations have for an origin the variations of the effective gravity, which could be considerable due to the changes of the radial pressure. It is possible that this mechanism can explain the phase relationship between the light curves and the radial velocity curves of the cepheids. It is possible that the radial velocities of the cepheids vary as a result of the variations of the radial pressure in an extended reversing layer, where the absorption lines are formed.

The question thus raised in the study of the RV Tauri stars makes it possible to re-examine from another viewpoint the properties of the cepheids.

REFERENCES

[1] F. Loukatskaya, *Variable Stars*, **9**, 57 (1952).
[2] A. M. Shulberg, *Variable Stars*, **9**, 256 (1953); *Publ. Astr. Obs. Univ. Odessa*, **3**, 249 (1953).
[3] I. G. Jdanova and V. P. Tsesevich, *Publ. Astr. Obs. Univ. Odessa*, **3**, 3 (1953).

V. PHENOMENA OF INSTABILITY IN BINARY SYSTEMS

17. THE INSTABILITY OF SUB-GIANTS IN CLOSE BINARY SYSTEMS

ZDENĚK KOPAL

Department of Astronomy, University of Manchester, Manchester, England

I should like to open my subject by attempting to answer the following question: how many parameters are necessary and sufficient for a complete specification of the form of both components in close binary systems? Their shape should, in principle, be specified by the nature of forces acting on their surfaces, and (provided that the free period non-radial oscillations of both components are short in comparison with that of their orbits) their distortion should be governed by the equilibrium theory of tides. The level surfaces of constant density then coincide with those of constant potential, and the boundary of zero density becomes a particular case of such equipotentials.

A complete theory of the form of such surfaces for stars of arbitrary structure has not so far been developed. If, however, the density concentration inside the components of close binary systems is so high that their gravitational potentials can be approximated by those of central mass-points, the gravitational potential acting on any surface point $P(r, \theta, \phi)$ of the component of mass $\mathbf{m}_1$ should be given simply by $G(\mathbf{m}_1/r)$; the potential arising from its mate of mass $\mathbf{m}_2$ should contribute a term $G(\mathbf{m}_2/r')$, where r' denotes the distance of P from the centre of gravity of the disturbing star, and G is the constant of gravitation. The centrifugal force due to rotation with an angular velocity ω about an axis perpendicular to the orbital plane should, moreover, give rise to a contribution $\frac{1}{2}\omega^2 D^2$, where D denotes the distance of P from the axis of rotation of the distorted component. The total potential W then becomes a sum of the three components just enumerated. In close binaries it seems, moreover, reasonable to identify ω^2 with the Keplerian angular velocity $G(\mathbf{m}_1 + \mathbf{m}_2)/R^3$, where R stands for the semi-major axis of the relative orbit of the two stars. Suppose now that we adopt the sum $\mathbf{m}_1 + \mathbf{m}_2$ as our unit of mass, R as the unit of length, and choose the unit of time in such a way that $G = 1$. If so, the desired equation of our equipotential surfaces can be shown to assume the neat form

$$C = \frac{2(1-q)}{r} + 2q\left\{\frac{1}{\sqrt{1-2\lambda r + r^2}} - \lambda r\right\} + r^2(1-\nu^2) + q^2, \tag{1}$$

where λ, μ, ν denote the direction cosines of an arbitrary radius vector r connecting P with the centre of mass of star 1, and

$$C = \frac{2RW}{G(m_1 + m_2)}, \qquad q = \frac{m_2}{m_1 + m_2}, \tag{2}$$

are non-dimensional constants.

The surfaces generated by setting $C =$ constant on the left-hand side of equation (1) can be appropriately referred to as the *Roche equipotentials*, and the C's themselves as *Roche constants*. If the latter are large, the corresponding equipotentials are known to consist of two separate ovals enclosing each one of the two mass-points, and differing but slightly from spheres. With diminishing values of C the ovals defined by (1) become increasingly elongated in the direction of the centre of gravity of the system until, for a certain critical value of $C = C_0$ (characteristic of each mass-ratio) [1], both ovals unite at a single point on the line joining the centres of the two stars. This limiting equipotential—the largest *closed* equipotential capable of containing the whole mass of the system—will hereafter be referred to as the *Roche limit*. For $C < C_0$ we can no longer regard the respective equipotentials as models of binary systems consisting of detached components, but for each value of $C \geqslant C_0$ equation (1) defines the surfaces of two distinct configurations which should describe the forms of centrally-condensed components of close binary systems, to a high degree of accuracy, *irrespective of their proximity or mass-ratio*. Therefore, the minimum number of parameters sufficient for a complete geometrical specification of *both* components in the close binary systems is *three*, and consists of the values of C_1, C_2, and q.* A properly determined trio of C_1, C_2 ($\geqslant C_0$) and q can describe the geometry of a system very much more simply and accurately than any number of artificial semi-axes of the individual components. Moreover, the quantities $C_{1,2}$ and q possess the additional advantage of a direct and simple physical meaning.

A determination of q from the spectroscopic data is sufficiently straightforward, and so is the determination of C from an analysis of the light curves [2]. Their values have recently been determined by the writer for all two-spectra eclipsing binaries of known light curves [3], and their discussion reveals that all systems possessing at least one (the more massive) component of the main sequence can be naturally divided into three groups of the following characteristics:

(a) *Stable Systems*. The volumes of both components are significantly

---

* Each one of the values of $C (\geqslant C_0)$ introduced in (1) defines, to be sure, a pair of such equipotentials for a given value of q, of which only the one enclosing the centre of gravity of the component under consideration is relevant.

smaller than their Roche limits, but their fractional dimensions and mass-ratios are such as to render the values of C for both components sensibly *equal* (in spite of the fact that the absolute values of the potential over free surfaces of the components vary from system to system by a factor in excess of ten). Both components do not deviate significantly from the main sequence in the H-R diagram, and obey statistical mass-luminosity and mass-radius relations.

(*b*) *Semi-Detached Systems*. The primary (more massive) components are significantly smaller than their Roche limits (and therefore dynamically stable), but the *secondaries* appear to fill *exactly* the largest closed equi-potentials capable of containing their whole mass (i.e. $C_2 = C_0$ within the limits of observational errors). Such components lie as a rule above the main sequence, and while their primaries conform to the same mass-luminosity and mass-radius relations as stars of group (*a*), the secondary components are mostly overluminous sub-giants.

A *complementary type* of semi-detached systems, with primaries at their Roche limits and secondaries well below it (i.e. characterized by $C_1 = C_0$ and $C_2 > C_0$) *seems conspicuous by its absence*.

(*c*) *Contact Systems*. Both components appear to fill the respective loops of their Roche limits and are, therefore, probably in actual contact. Both stars lie (though not very closely) along the main sequence, but show—individually or statistically—no vestige of any relation between mass and radius or luminosity.

A schematic representation of the geometry of these three types is shown on the accompanying Fig. 1, drawn to scale for a mass-ratio $m_2/m_1 = 0.6$. The question of classification of eclipsing binary systems was discussed recently at a meeting of the I.A.U. Commission 27 (Variable Stars) in connexion with the continued practice by Kukarkin and Parenago to use Algol, β Lyr, and W UMa as prototypes of such variables in their well-known *Catalogue*. It would appear now that this older system of classification has little to recommend it except tradition. The most common type of eclipsing binaries—the main sequence (stable) systems of our group (*a*) —is not recognized by it (β Aur or U Oph could be regarded as suitable prototypes). Algol itself is a typical representative of variables of our group (*b*) (semi-detached systems), while W UMa does (and β Lyr may) belong to our group (*c*). Algol and W UMa thus could be regarded as genuine prototypes of their groups, but β Lyr represents too peculiar and unique a system to be suited for the prototype of any group of common eclipsing systems. Its use in this role so far could be justified only on histori-cal grounds, and the principal distinguishing feature between 'Algol' and

'β Lyr stars'—namely, the presence or absence of the effects of photometric ellipticity between minima—is only a matter of observational precision.

From the dynamical point of view, systems of group (*a*) should be regarded as stable unless forces other than gravitational or centrifugal are

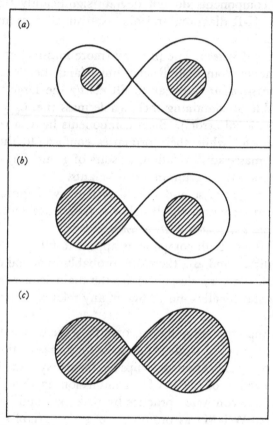

Fig. 1. A schematic view of the three principal types of close binary systems: (*a*) main sequence (stable) systems, (*b*) semi-detached systems, (*c*) contact binaries. The diagrams are drawn to scale for a mass-ratio of $m_2/m_1 = 0.6$.

operative. Instability phenomena are, in principle, likely to occur among systems of group (*b*) as well as (*c*). In the present communication we shall not, however, be concerned with contact systems and propose to confine our attention to semi-detached systems whose existence is, in many respects, particularly thought-provoking. For, consider a group of well-known eclipsing systems of group (*b*), compiled in Table 1, that have been selected to demonstrate the closeness with which their secondary components cling to their Roche limits. Column 3 of this tabulation contains the values of spectroscopic mass-ratios m_2/m_1 of both components together with their observational uncertainty, and column 4 gives the diametral semi-axes of their corresponding Roche limits, taken from a recent investigation by the writer [4], their uncertainty as given being due solely to that of the mass-ratios. The last column lists then the actual diametral semi-axes of the secondary components in these systems, as deduced from an analysis of the light curves. An inspection of the data in columns 4 and 5 demonstrates that the theoretical and observed semi-axes are significantly the same in every single case, and their coincidence tends to become the closer, the smaller the uncertainty of the underlying observational data. Several other well-known systems could, moreover, be added to augment our list.

Table 1. *Fractional Dimensions of Secondary (Contact) Components in Semi-Detached Binary Systems*

Star	Spectral type	m_2/m_1	$(r_2)_{comp.}$	$(r_2)_{obs.}$
RT And B	K1	$0·65 \pm 0·03$	$0·333 \pm 0·007$	$0·325 \pm 0·004$
U Cep B	gG8	$0·49 \pm$	$0·31 \pm$	$0·31 \pm 0·01$
U CrB B	gG0	$0·38 \pm 0·01$	$0·290 \pm 0·003$	$0·28 \pm 0·01$
u Her B	B7	$0·35 \pm 0·02$	$0·285 \pm 0·004$	$0·287 \pm 0·003$
V Pup B	B3	$0·58 \pm 0·02$	$0·324 \pm 0·004$	$0·327 \pm 0·004$
U Sge B	gG6	$0·30 \pm 0·02$	$0·272 \pm 0·005$	$0·278 \pm 0·003$
V 356 Sgr B	A2	$0·38 \pm 0·03$	$0·292 \pm 0·007$	$0·28 \pm 0·01$
μ^1 Sco B	B6	$0·66 \pm 0·02$	$0·337 \pm 0·004$	$0·347 \pm 0·004$
TX UMa B	gG4	$0·30 \pm 0·02$	$0·272 \pm 0·004$	$0·277 \pm 0·001$
Z Vul B	(A2)	$0·45 \pm 0·02$	$0·303 \pm 0·003$	$0·301 \pm 0·002$
RS Vul B	(F4)	$0·31 \pm 0·03$	$0·274 \pm 0·007$	$0·26 \pm 0·01$

What is the significance of this clustering of secondary components in systems of this type around their Roche limits? The fact is certainly not the result of chance, for the probability of so peculiar a random distribution of fractional dimensions in so large a sample is negligibly small. It indicates rather that these stars have reached their limits by non-equilibrium processes. If they were contracting, there is no reason why any number of them should cluster around the Roche limit, but if they expand, the reason for such a clustering becomes compelling, since no larger *closed* equipotential exists which would contain their whole mass. Therefore, the growth of an expanding component of a close binary system is bound to be *arrested* at its Roche limit, and if this tendency is latent in such stars as a group, they should indeed be expected to cluster around this limit.

The observed facts just discussed can, therefore, scarcely be accounted for otherwise than by a hypothesis that the *sub-giant components in close binary systems are secularly expanding* [5]. Once the maximum distension permissible on dynamical grounds has been attained, however, a continuing tendency to expand is bound to bring about a *secular loss of mass*, by the streaming of material out of the conical end of the critical equipotential (at which the previously closed surface begins to open up). What should be the kinematic behaviour of the ejected matter once it has left the secondary component? If we ignore minor perturbations arising from the finite degree of central condensation of both components and retain the same system of units as used previously, the motion of a gas particle of negligible mass in the gravitational dipole field generated by the finite masses $\mathbf{m_1}$ and $\mathbf{m_2}$ (separated by constant distance), and referred to a rotating rectangular frame of reference whose x-axis coincides with the line joining $\mathbf{m_1}$ and $\mathbf{m_2}$ and whose y-axis lies in the plane of the orbit, should be governed by the well-known equations of the restricted problem of three bodies. Moreover, if we limit our attention to orbits in the orbital (*xy*-) plane, the respective equations assume the form

$$\frac{d^2x}{dt^2} - 2\frac{dy}{dt} = \frac{\partial U}{\partial x},$$

$$\frac{d^2y}{dt^2} + 2\frac{dx}{dt} = \frac{\partial U}{\partial y},$$

(3)

where

$$U = \tfrac{1}{2}(x^2 + y^2) + \frac{1-q}{r} + \frac{q}{r'}$$

(4)

stands for the potential energy of our system. It also, incidentally, represents the rectangular–co-ordinate version of equation (1) and, at the same time, a Jacobian surface of zero velocity [6] of the moving gas particle.

The actual form of trajectories governed by the foregoing equations depends, of course, on the initial conditions of escape, and these are not yet known with any uniqueness. The locus of ejection is no doubt the conical point of the critical equipotential enclosing the less massive component (see again Fig. 1 (*b*)), identical in fact with the inner Lagrangian point L_1 at which both the velocity and acceleration of any mass particle vanish in the *xy*-plane. If this particle were subject to no exterior force, it would remain permanently at rest there (relative to our moving frame of reference). Conversely, an application of exterior force would require less energy to remove the particle from L_1 than from any other point on the star's surface. Concerning the nature of this force, in what follows we wish to explore the consequences of a hypothesis that the ejection is caused by a difference between the actual angular velocity ω_E of rotation at the equator of the secondary component in semi-detached binary systems, and

128

their Keplerian angular velocity ω_K. If $\omega_E \neq \omega_K$, matter should keep moving along the equator in the xy-plane (clockwise if $\omega_E < \omega_K$, counterclockwise if the converse is true) and, on arriving at L_1, should be ejected towards the primary (more massive) component in the direction subtending an angle with the x-axis equal to that of the osculating cone with vertex at L_1 [7]. The velocity of ejection should depend on the magnitude of $|\omega_E - \omega_K|$ and may be arbitrary within wide limits.

In order to explore the topological properties of gas streams ejected under these conditions, numerical integrations of several hundred trajectories have recently been undertaken at Manchester, for different mass-

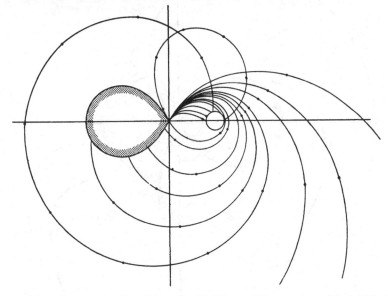

Fig. 2. Trajectories of particles ejected from L_1 with thirteen different values of the initial velocity V_0 ranging from 0·5 to 3·0, for the case of $\omega_E > \omega_K$ (direct orbits). The values of V_0 are given in Table 2. The mass-ratio m_2/m_1 is 1·0.

ratios and diverse velocities of ejection when $\omega_E \lesssim \omega_K$. These integrations have been carried out with the aid of the University of Manchester's Electronic Computers (Marks II and III) and have been programmed by my colleague R. A. Brooker of the Computing Machine Laboratory, assisted by Miss Vera Hewison of the Department of Astronomy (who is also responsible for graphical presentation of all results to be given below). The technical aspects of automatization of the differential equations of the problem of three bodies will be described by Brooker elsewhere. The aim of the present communication will be to survey the principal astronomical results obtained so far, and to give a preliminary discussion of their significance.

The accompanying Figs. 2–7 show graphical representations of the single-parameter families of the ejection orbits from L_1 for mass-ratios $m_2/m_1 = 1\cdot0$, $0\cdot8$, and $0\cdot6$, and different sets of initial velocities, the two cases $\omega_E \gtrless \omega_K$ being treated separately. On each diagram, the centre of gravity of the system is taken as the origin of co-ordinates, and the outline

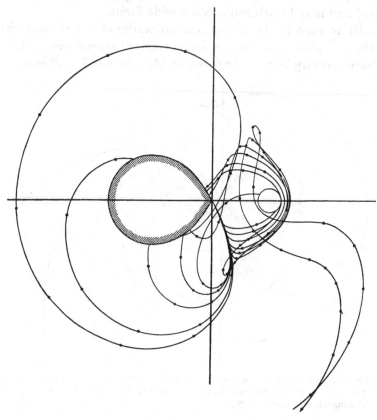

Fig. 3. Trajectories of particles ejected from L_1 with thirteen different values of the initial velocity V_0 ranging from $0\cdot5$ to $1\cdot6$, for the case of $\omega_E < \omega_K$ (retrograde orbits). The mass-ratio m_2/m_1 is $1\cdot0$.

of the secondary's equator is drawn to scale for the respective mass-ratio. The circle of radius $0\cdot1$ enclosing the primary's centre of gravity, however, merely represents a limit inside which we began getting in trouble with the scale factors of our machine integrations, rather than any anticipated size or shape of the primary component (which can be arbitrary inside its own Roche limit). Filled circles on each trajectory represent points attained by the moving particles in equal time intervals. A list of the initial velocities of all trajectories plotted on Figs. 2–7 is given in Table 2.* A great many more integrations than those shown here have been performed at Manchester to date,† but their present selection should illustrate sufficiently the main topological properties of ejection orbits (under the envisaged conditions) in the equatorial plane. Considerable work has also been done

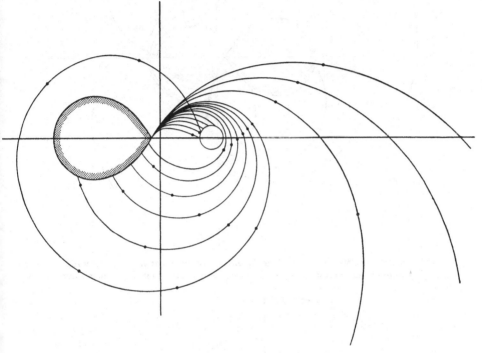

Fig. 4. Trajectories of particles ejected from L_1 with fourteen different values of the initial velocity V_0 ranging from 0·5 to 3·5, for the case of $\omega_E > \omega_K$ (direct orbits). The mass-ratio $\mathbf{m_2}/\mathbf{m_1}$ is 0·8.

on trajectories corresponding to the mass-ratio 0·4, but dynamical conditions in this latter case have proved to be considerably more complex, and a fuller presentation of the results is being postponed for a later occasion. We cannot, however, forego exhibiting on Fig. 8 at least one trajectory of this family (corresponding to the initial velocity of $v_0 = -0·7$)‡ on account of its re-entrant character.

* The unit of velocity, consistent with our previous practice, becomes $\sqrt{G(\mathbf{m_1}+\mathbf{m_2})/R}$ cm./sec.
† It is estimated that the machine solutions made so far are equivalent to at least 50,000 man-hours with an ordinary desk-type computing machine.
‡ In what follows, the negative sign will be used to denote velocities of ejection for $\omega_E < \omega_K$ (i.e. when y is negative), giving rise to retrograde orbits.

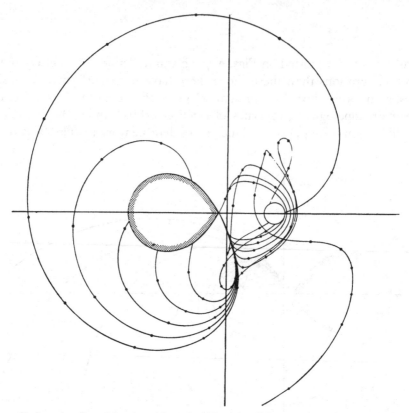

Fig. 5. Trajectories of particles ejected from L_1 with twelve different values of the initial velocity V_0 ranging from 0·5 to 1·6, for the case of $\omega_E < \omega_K$ (retrograde orbits). The mass-ratio $\mathbf{m_2/m_1}$ is 0·8.

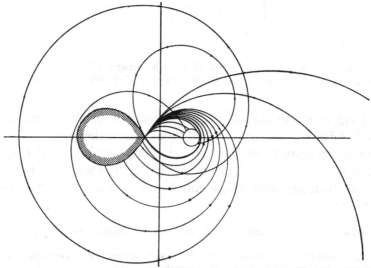

Fig. 6. Trajectories of particles ejected from L_1 with thirteen different values of the initial velocity V_0 ranging from 0·5 to 4·0, for the case of $\omega_E > \omega_K$ (direct orbits). The mass-ratio $\mathbf{m_2/m_1}$ is 0·6.

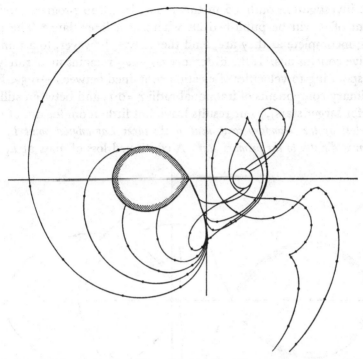

Fig. 7. Trajectories of particles ejected from L_1 with nine different values of the initial velocity V_0 ranging from 0·5 to 1·5, for the case of $\omega_E < \omega_K$ (retrograde orbits). The mass-ratio $\mathbf{m_2}/\mathbf{m_1}$ is 0·6.

Table 2. *The Initial Velocities (in Units of $\sqrt{G(\mathbf{m_1}+\mathbf{m_2})/R}$ cm./sec.) of the Ejection Orbits Plotted in Figures 2–7.*

Figure: 2	4	6	3	5	7
Direct orbits: $\omega_E > \omega_K$			Retrograde orbits: $\omega_E < \omega_K$		
$\mathbf{m_2}/\mathbf{m_1}=1\cdot0$	0·8	0·6	1·0	0·8	0·6
0·5	0·5	0·5	0·5	0·5	0·5
0·7	0·7	0·8	0·6	0·6	0·8
1·0	0·9	1·1	0·7	0·7	0·9
1·3	1·1	1·4	0·8	0·8	1·0
1·4	1·3	1·5	0·9	0·9	1·1
1·5	1·5	1·6	1·0	1·0	1·2
1·7	1·6	1·7	1·05	1·1	1·3
1·8	1·7	1·8	1·15	1·2	1·4
1·9	1·8	1·9	1·2	1·3	1·5
2·0	1·9	2·0	1·3	1·4	—
2·25	2·0	2·5	1·4	1·5	—
2·5	2·5	3·0	1·5	1·6	—
3·0	3·0	4·0	1·6	—	—
—	3·5	—	—	—	—

The investigation outlined in this report is still in progress, and a full account of it will be published elsewhere at a later date.* The present results, incomplete as they are, lend themselves, however, to a number of tentative conclusions. If the difference $\omega_E - \omega_K$ remains moderately small (corresponding to velocities of ejection contained between $-0.5 < V_0 < 1.5$ for primary components of fractional radii $r = 0.1$, and between still wider limits for larger stars),† our results leave but little room for doubt that *all matter lost by the secondary component at the inner Lagrangian point L_1 will be transferred directly to the primary star.* A continued loss of mass at L_1 due to

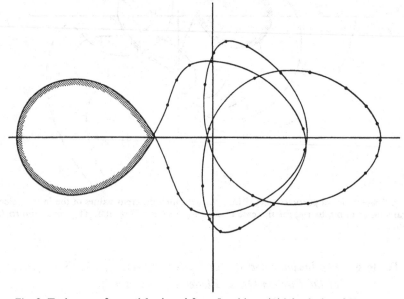

Fig. 8. Trajectory of a particle ejected from L_1 with an initial velocity of $V_0 = 0.7$ in the case of $\omega_E < \omega_K$ (retrograde orbits), and for a mass-ratio of $m_2/m_1 = 0.4$.

secular expansion of the secondary component is, therefore, in time bound to keep increasing the disparity in masses between the two components of the same pair. Now close eclipsing systems with sub-giant secondaries have long been known to exhibit abnormally large mass-ratios—a fact discussed particularly by Struve [8]. As most (if not all) such companions have been found to possess fractional dimensions coinciding with their Roche limits, there remains but little room for doubt that the relative smallness of their present masses is but the consequence of a secular transfer of mass from the

* We shall, however, be pleased to furnish particular results, in advance of publication, to any investigator who may request them.
† It should be stressed that, in most binary systems, the actual velocities of ejection are likely to lie within these limits.

secondary to the primary components by 'gravitational pipelines' shown on our Figs. 2–7.

If the velocity of ejection becomes higher (its exact limit depending on the size of the primary component), the mass ejected at L_1 may circumnavigate the primary component and be intercepted by the secondary

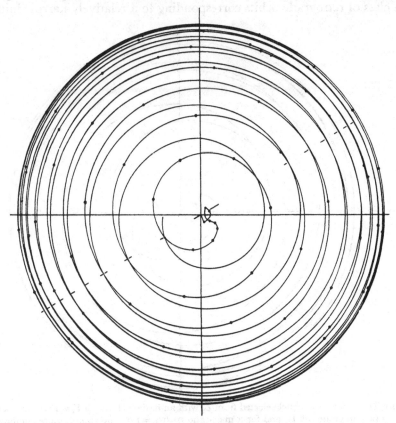

Fig. 9. Trajectory of a particle ejected from L_1 with an initial velocity of $V_0 = 0 \cdot 9$ in the case of $\omega_E < \omega_K$ (a retrograde orbit), and for a mass-ratio $m_2/m_1 = 1 \cdot 0$. The scale is more compressed than that of Figs. 2–7.

anywhere along its equator (cf. again Figs. 2–7), or, with further increase of ejection speed, circumnavigate the primary *and* secondary until falling at last on the primary star. It is not until still higher velocities are attained— velocities much higher than the radial velocities of any gas streams actually observed in eclipsing variables (with the possible exception of β Lyrae, or of the Wolf-Rayet and other anomalous binaries)—that matter ejected by the secondary component at L_1 can actually spiral out in the equatorial plane and be lost to the system. In fact, one of the principal results of the present investigation is the full realization of the difficulty with which any matter can be permanently expelled from a close binary system.*

If $\omega_E > \omega_K$, the foregoing description covers broadly any dynamical contingency which may arise, but for $\omega_E < \omega_K$ an additional interesting possibility should be mentioned in this connexion: namely, the existence of a class of retrograde orbits corresponding to a relatively narrow initial-

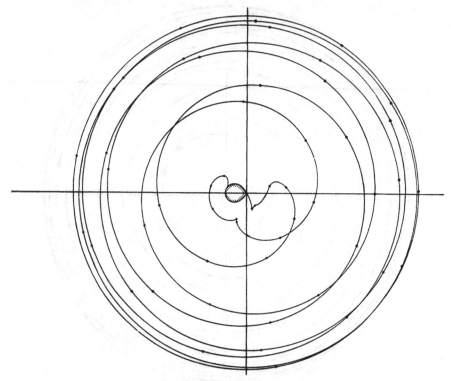

Fig 10. Trajectory of a particle ejected from L_1 with an initial velocity of $V_0 = 1\cdot0$ in the case of $\omega_E < \omega_K$ (a retrograde orbit), and for a mass-ratio $m_2/m_1 = 1\cdot0$. The scale is more compressed than that of Figs. 2–7. Fig. 3 should be consulted for the details of the inner part of the trajectory.

velocity range around $V_0 = -1\cdot0$. Our numerical integrations show (cf. Figs. 3, 5 and 7) that the particles departing from L_1 with such a velocity may, after circumnavigating the primary component (and describing cusps in the neighbourhood of the Lagrangian equilateral points L_4 and L_5), pass through the gap between the two stars and recede thereafter to a considerable distance from the system. In two cases (corresponding to the mass-ratio $1\cdot0$ and the ejection velocities $V_0 = -0\cdot9$ and $-1\cdot0$), we have

* The opening-up of the corresponding Jacobian surface of zero velocity is a necessary, but not sufficient, condition for such an escape.

followed the orbits of such particles for many thousands of steps (requiring integrations which took hours to perform even at electronic speeds).

The outcome, plotted on Figs. 9 and 10 on a much reduced scale, reveals that the recession of a particle from its parent binary system does not continue indefinitely, but possesses an upper limit (of radius about ten times as large as the radius of the orbit of the two finite bodies), which our particle will approach asymptotically before it will eventually spiral inward to end its motion by falling on one of the two components. The asymptotic nature of the orbit (cf., in particular, Fig. 9) suggests that a continuous stream of particles moving along it may lead to the maintenance of a distinct gas ring, encircling the whole binary system at a considerable distance from it and (as indicated by the time-dots) rotating with respect to it with almost constant angular velocity. Now the existence of gaseous rings of this nature, encircling certain eclipsing systems, has indeed been reported by several investigators. Whether or not these are dynamically related to the secondary component by particle orbits discussed in this paper, however, remains to be settled by future investigations.

REFERENCES

[1] For their tabulation cf. Kopal, Z., *Jodrell Bank Ann.* **1**, 37 (1954) (Table V, col. 2).
[2] Details of this process are being postponed for subsequent publication.
[3] The material at the basis of this study is summarized in Table V of the Draft Report prepared by the writer on behalf of Commission 42 of the I.A.U. for the 9th General Assembly in Dublin, and will be published in the *I.A.U. Transactions*, vol. **9**.
[4] Cf. Kopal, Z., *Jodrell Bank Ann.* **1**, 37 (1954) (Table I, col. 7).
[5] This conclusion, announced by the writer at the sixth International Astrophysical Colloquium at Liège in July 1954 (cf. *Communications présentées au sixième Colloque International d'Astrophysique*, Liège, 1955, pp. 684–5), was independently arrived at by J. A. Crawford [*Ap.J.* **121**, 71 (1955)] on the basis of a statistical study of completely different material.
[6] For fuller details cf., e.g., F. R. Moulton, *An Introduction to Celestial Mechanics* (6th ed., New York, 1939) chapter VIII, sec. 154.
[7] These angles have been tabulated for different mass-ratios by Kopal in *Jodrell Bank Ann.* **1**, 37 (1954) (Table III).
[8] Struve, O., *Ann. d'Ap.* **11**, 117 (1948); also *Harvard Centennial Symposia* (*Harv. Obs. Mono.* No. 7, Cambridge, 1948), pp. 211–30.

18. PHENOMENA OF NON-STABILITY IN CLOSE BINARY SYSTEMS

D. J. MARTYNOV

Engelhardt Observatory, Kasan, U.S.S.R.

(1) The non-stability of the components in close binary systems has been established in a number of systems, mainly among the eclipsing variables. It cannot be regarded as an unusual phenomenon.

Several types of non-stability may be pointed out:

(*a*) Physical variability of one or two of the components as, for example, in the following systems:

Star	Spectral types	Remarks
S Dor	P Cyg	
AR Pav	W+P+cF	
VV Cep	B3+gM2	
32 Cyg	B8+cK5	
RX Cas	gA5e+gG3	Pulsation is observed with a period greatly exceeding orbital period
UX Mon	A+G0–2 (III–IV)	
U Peg	F3+F3	Variations of brightness are particularly large in ultra-violet
VW Cep	dG5+dK1	
UX UMa	B3 (sd)	Same as U Peg

Non-stability of this kind is obviously a feature not confined to binary systems.

(*b*) The presence of emission bands or lines in the spectrum in every or some phases of the orbital motion as, for example, in the case of the eclipsing variables:

Star	Spectral types
V444 Cyg	WN5+O6
CQ Cep	WN6
CX Cep	WN5
V729 Cyg	Of+O9

The same phenomenon is observed in the case of HD 228786 and some other spectroscopic binaries having Wolf-Rayet components. It is suggested that the absorption spectra of all Wolf-Rayet stars originates in the satellite of type O. There exists a still more radical opinion that all Wolf-Rayet stars are binaries. Thus, non-stability of the Wolf-Rayet type may be typical for close binary systems.

Closely similar to these stars are systems in which the intensity of the emission characteristics, if they are observed in all phases of the orbital motion, varies with the phase. Such objects are:

Star	Spectral types	Remarks
CQ Cep	WN6	He II λ 4686 is most intense at conjunctions
*β Lyr	B9+F	Same as CQ Cep
V367 Cyg	B8–9 (?) or F0 ($\pm$2) III–V	
RZ Oph	F3 I+gK5p	
*RX Cas	gA5e+gG3	
SX Cas	cA6+G6	
UX Mon	A+G0–G2 (III–IV)	

These characteristics are manifested by other systems only in certain phases of the orbital cycle as, for instance, in:

Star	Spectral types	Remarks
*β Per	B8+G	The possible ejections observed at the maximum phase of the eclipse appear only rarely
*U Sge	B9n+gG2	The same as β Per
*RW Tau	B9e+K0	The same, but more stable
*AR Lac	G5+K0	
W UMa	F8p+F8p	H and K are in emission .
RZ Cnc	K2+K5	H and K are in emission in the K2 component
YY Gem	dK6+dK6	H, K and hydrogen lines are in emission in both components
UX UMa	B3 (sd)	Hβ is in emission at certain phases
31 Cyg	B+gK5	Turbulent motions exist in the atmosphere of the K5 component

(c) This non-stability manifests itself in sharp differences between the spectroscopic and photometric elements of the system. The discrepancies are caused by gaseous streams in the regions where the absorption lines originate. These lines.are either displaced, due to the Doppler effect, or have profiles distorted owing to the same cause. Such phenomena are found in the following systems:

Star	Spectral types
RZ Sct	B2
*U Cep	B8+gG2
*RX Cas	gA5e+gG3
SX Cas	cA6+G6
UX Mon	A+G0–G2 (III–IV)

(d) This non-stability manifests itself in an unsteadiness of the period of orbital motion. In some cases the observed times of light minima cannot be represented either by a linear ephemeris or by any other formula, if the latter is applied for an interval of time different from that for which it was deduced. Such non-stability is represented by a number of systems not enumerated here. It should, however, be pointed out that a number of the systems already mentioned have inconstant photometric periods (such systems are marked with an asterisk).

(2) As is well known, stars of quite different physical properties (revealed by their spectra and absolute magnitudes) are found in close pairs. Certain combinations are never met with as, for example, a giant together with a dwarf of late spectral type. This might easily be understood as a result of observational selection. Therefore, the non-stability of type (*a*) is actually not typical for binary systems. Contrariwise, the origin of the non-stability of type (*b*) is facilitated by the existence of a companion near a given star, or is caused by the tidal action of the secondary. In the above lists we meet representatives of about all types of stars from the spectrum-luminosity diagram. This means that normal and hot giants, B- and A-type stars, ordinary dwarfs, sub-dwarfs and sub-giants are found among unstable stars. The features of non-stability (the presence of a gaseous ring) are so weak in the systems of U Sge, RW Tau and possibly β Per, that they become visible only when the total, or nearly total, eclipses cut off the light of the photosphere of the bright component. If no eclipses should occur, we would know nothing regarding such types of non-stability, especially since they are transient phenomena. At the same time these phenomena are apparently confined to close binaries. They may exist in a number of spectroscopic binaries, but remain unobserved in the absence of eclipses.

The eclipsing variable DQ Her—a former nova—is an example of another form of non-stability. This non-stability was, perhaps, the cause of the origin of this extremely close binary system, whose spectrum even now manifests features of non-stability. We do not know, however, how long such features will survive.

(3) We have at present for the explanation of the non-stability in close binary systems the so-called Kuiper-Struve mechanism. At a sufficiently small separation of the components, the geometrical dimensions of one or both may surpass the limits of the internal or external equipotential surfaces. There will take place in such a case either the exchange of matter between the components along the equipotential surfaces, or an ejection of matter from the system through the second Lagrangian point, on the condition that the total energy of the particle will exceed some critical value h_{L_2}. If neither of the components of such a 'contact binary star' surpasses the limits of the outer equipotential surface, ejection of matter might, however, take place owing to the thermal non-stability of one component. That is, macroscopic motions in the stellar atmosphere might lead to an

ejection of matter possessing sufficient kinetic energy beyond the limits of the internal equipotential surface. In this case the phenomenon, which would have no consequences for a single star, would lead in a double system to a loss of mass and of rotational momentum. Spectroscopic effects of such ejection have been discovered by O. Struve and others since 1941 in a number of photometric binary systems.

As concerns photometric effects, the asymmetry of light curves observed in some eclipsing variables might be explained, as it was first by Mergentaler in 1950, in terms of streams of gaseous material. Indeed, large masses of cool gas actually cover the photosphere of the star, decreasing its surface brightness. In so far as the gaseous stream is asymmetric, the steepness of the light curve and the total brightness of the system before the eclipse and after will be quite different.

Unfortunately, the quantitative deductions of these effects made by Mergentaler, Dadaev, Sofronicky, and Svetchnikov were carried out by these authors under extremely simplified conditions, because of difficulties of a mathematical nature. These calculations show only that the expected effects will be of the same order as the ones observed. A more detailed mathematical and physical analysis of this phenomenon is extremely desirable for the elimination of the effects of asymmetry in the observed light curves. All the modern methods for the solution of the light curves of eclipsing variables have attained a high degree of perfection, but without taking into account this asymmetry.

Svetchnikov calculated also the intensity of the emission lines that might be observed in the spectrum of the star in the presence of gas streams, and found them (also under simplified assumptions) to correspond with the observed intensity of the emission lines.

All these results render the hypothesis of the ejection from the contact binaries trustworthy, but not yet proven by far.

(4) However, the hypothesis of ejection is supported in another way, since it may be used for the explanation of irregular changes in the length of the orbital period of a binary system. As has been said before, such variations of the period are a common phenomenon in eclipsing variables.

Celestial mechanics points out only two sources of the variation of the photometric period of a binary system: motion around a third body and motion of the apsidial line in the case of eccentric orbit. These two cases are represented in the eclipsing variables by some reliable examples.

The motion around a third body is observed in the cases of β Per, RT Per(?), and possibly SW Cyg; the motion of the apsidial line is observed in the binary systems YY Cyg, RU Mon, GL Car, V526 Sgr, and YY Sgr.

Both motion of the apsidal line and motion around a third body, the theories of which give a number of additional terms in the expression for the epochs of photometric conjunctions, are manifested by periodic (usually long-periodic) terms in the epochs of minima. Meanwhile, irregular variations of the epochs of minima of eclipsing variables, however, are quite beyond doubt in a number of cases. The circumstance that the photometric period does not coincide with the sidereal period of orbital motion cannot explain this phenomenon, because the observed differences even in the case of the libration of ellipsoidal components (if librations are possible in a system of gaseous stars), must also be regular and not erratic.

It is clear that the fluctuations of the photometric period in a binary system correspond to the actual fluctuations of the sidereal period of orbital motion, which is possible only if changes of the rotational momentum of the system are taking place. The hypothesis of gaseous streams makes such fluctuations admissible. Individual spontaneous fluctuations of the period might be explained, as was done by Wood, by transitory ejections of very large masses of matter from a star and a system. Whereas a continuous loss of mass by a system must lead to a secular increase of the period corresponding to the law $a(\mathbf{m}_1 + \mathbf{m}_2) = \text{constant}$, a one-sided ejection (depending upon its direction) may cause the period to become longer or shorter. As is seen from the calculations by Wood and by Svetchnikov, the observed fluctuations of the period in RZ Cas, U Cep, AO Cas, AR Lac and U Sge require an ejection of the order of 10^{-7} or 10^{-6} solar masses. If we remember that the masses ejected at nova outbursts exceed the aforementioned values only by one order of magnitude, it seems doubtful that ejections of masses of the order of 10^{-7} or 10^{-6} solar masses might pass unnoticed for stars under frequent observation. But it must also be remembered that the above ejections will explain the observed spontaneous changes in the periods, if they represent a series of similarly directed ejections that continue for a considerable interval of time, for example, a year. The photometric effects calculated by Svetchnikov (though under simplified conditions) will then be found to be sufficiently weak. However, single large changes of the period require the simultaneous ejection of large masses, which could not escape the attention of the observers.

As was shown by the author some time ago, the phenomenon of tidal friction might, under certain conditions, facilitate the mutual approach of the components in binary systems, if the viscosity of matter and of radiation are considerable in a star. Ejections of matter, if directed in a suitable way, might assist this approach. Thus, the components may approach each other so closely that the ejection of matter from them will become extremely intense. But a similar stage of strong non-stability cannot be of long duration, because the law $a(m_1 + m_2) = $ constant will lead to an increase of the distance between the components. If tidal friction is absent, the stage of non-stability of a star will be of very long duration (of the order of millions of years) only if the change of the structure of a component of the system will lead to a progressive increase in its dimensions.

Thus we find a number of mechanisms that make the components of close binaries non-stable in the course of considerable intervals of time. This makes their evolution quite different from the evolution of a single star.

BIBLIOGRAPHY

A. N. Dadaev, *Pulkovo Bull.* **19**, No. 5, 31 (1954).
Z. Kopal, *Jodrell Bank Ann.* **1**, 37 (1954).
G. P. Kuiper, *Ap.J.* **93**, 133 (1941).
D. J. Martynov, *Publ. Engelhardt Obs.* No. 25 (1948).
J. Mergentaler, *Contr. Wroclaw Astr. Obs.* No. 4 (1950).
A. V. Sofronicky, *Pulkovo Bull.* **19**, No. 4, 1 (1953).
A. V. Svetchnikov, *Variable Stars*, **10**, 262 (1955).
F. B. Wood, *Ap.J.* **112**, 196 (1950).

19. PHOTOMETRIC EVIDENCE OF INSTABILITY IN ECLIPSING SYSTEMS

FRANK BRADSHAW WOOD

Flower and Cook Observatories, Philadelphia, Pennsylvania, U.S.A.

Many eclipsing variables exhibit characteristics that indicate lack of stability. Physically, these systems range from certain short-period dwarfs, which show irregular brightness fluctuations unexplainable by any normal eclipse hypothesis, to systems such as AO Cassiopeiae: class O super-giants whose periods, velocity curves, and light curves have shown remarkable variations. Included are systems having one Wolf-Rayet component and probably one old nova. Many eclipsing binaries show erratic changes of period that are difficult to explain on any concept of a stable system. Indeed, when we consider the physical conditions which must prevail when two stars are located with their surfaces only a few hundred thousand miles from each other, perhaps we should be surprised that the irregularities are not larger.

The general consideration of signs of non-stability shown by eclipsing systems is thus an extensive field. A great deal of evidence is now available, but some of it is unpublished, and the rest is rather widely scattered in the literature. The effort here will be to present a summary of photometric evidence that indicates instability is present in certain types of eclipsing systems.

Perhaps the first sound observational evidence that indicated that eclipsing systems could not always be represented by stable models was the visual observations of R Canis Majoris made in the years 1896–9 by Pickering[1] and by Wendell[2]. On certain nights in 1898, Wendell's observations showed clearly a 'hump', or region of increased brightness, at the end of primary minimum. This has been discussed in some detail by Dugan[3], who concluded that the hump undoubtedly had real significance, but could scarcely be a permanent feature of the curve. Indeed, observations Wendell made only a week later fell about 0·13 magnitude below the hump and agreed well with observations at maximum light elsewhere on the curve. Observations on one night in 1899 show a strikingly similar hump of slightly smaller amplitude. Pickering's mean curve shows a similar phenomenon. The scatter in Pickering's individual observations is larger than that in Wendell's, but he himself seemed fully convinced of the reality of the hump and speculated as to its cause. However, it is not found every cycle at this particular phase, and an analysis of Pickering's observations shows that it arises chiefly from observations made on two particular nights. Apparently by chance, these were nights on which Wendell also was observing the system and these are two of the nights in which the hump was clearly present in Wendell's observations. The degree of confirmation is thus rather strong.

No such phenomenon is found in Dugan's observations (1916–23) nor in any of the extensive later work. Dugan pointed out that the dispersion of Wendell's observations is much greater in the region of the hump than in the rest of the curve. This seems to reflect differences from night to night rather than increased scatter on a given night.

The observations by Pickering and by Wendell were made with polarizing photometers. In the hands of an experienced observer, this instrument is remarkably free from systematic error. The observer estimates when two nearby star images are equally bright, and is not aware of the actual values of the measures until he reduces his observations. Provision is made for interchanging the apparent positions of comparison and variable in the eyepiece and for reversing the optical parts of the photometer. Thus are eliminated the psychophysiological effects that caused strange phenomena to be reported by observers making visual estimates of brightness. We must conclude with Dugan that the hump was a real but non-permanent feature of the light curve.

Since several series of observations since 1916 have shown no trace of the hump, it seems pertinent to ask if evidence exists which indicated other unusual occurrences in the years preceding 1916. Two different kinds of evidence exist.

The first is the spectrographic work of Jordan [4] based on radial velocity measures made in the years 1908–12. Two features stand out in these observations. One is the internal inconsistency of the spectrographic observations in the individual years; the other is the large disagreement between the photometric and spectrographic orbit solutions. It seems reasonable to assume that these features are not disconnected and that whatever physical cause is responsible for the variation in the radial velocity measures is also responsible for this disagreement with the photometric elements.

In Hardie's [5] work on U Cephei a similar problem is discussed, and an answer is suggested that is now generally accepted: namely, that errors in estimating the central positions of the lines are caused by strong line asymmetry that varies with phase. The cause of the asymmetry is assumed to be absorption in circumstellar streams or shells of gas. But there is a significant difference between the radial velocity measures of R Canis Majoris and those of U Cephei. The U Cephei measures have always indicated a large—although not constant—eccentricity, in contradiction to the photometric evidence. This is not the case with R Canis Majoris, for later radial velocity observations by Sitterly[6] (1929–31) and by Struve and Smith[7] (1948–9) do not require the large eccentricity indicated by Jordan's observations. Indeed, both later series indicate that the orbit may well be circular.

Table 1. *Photometric Elements of R Canis Majoris*

	Wendell (vis 1898)	Dugan (vis 1928)	Wood (pe 1939)
$1-\lambda_1$	0·081	0·028	0·035
$1-\lambda_2$	0·426	0·373	0·405
L_b	0·834	0·943	0·928
L_f	0·166	0·057	0·072

A value of $\alpha_0 = 0.489$ was used in computing the L's.

The second unusual occurrence at about the epoch of Jordan's observations was a large, and apparently relatively sudden, period change. Although the period of R Canis Majoris had remained constant or nearly so over the preceding thirty years, a change of nearly a second occurred very close to the time when Jordan's spectrographic observations were showing fluctuations in velocity.

As a working explanation, we might assume the presence of the hump to be evidence of a developing unstable condition—a condition that terminated in ejection of material which distorted Jordan's spectrographic measures and which caused the period change by means of a physical mechanism suggested previously[8]. A further shortening of the period and the continued existence of discrepancies between the two later sets of spectrographic elements suggest that the final story may not yet have been told. Neither the photometric nor the spectrographic observations of the system since 1916 show the marked irregularities existing before that date.

If a catastrophic phenomenon of this kind did occur shortly before Dugan's observations so that the ejected material was still in the general neighbourhood of the stars, we might expect to notice some effect upon his light curve. A comparison of Dugan's observations with those of Wendell and with later photo-electric observations[9] shows serious discrepancies in the observed depths of minima. This is illustrated by the elements listed in Table 1. Even though the photo-electric observations were not available at the time of his study, Dugan was well aware of the difference between the

146

two sets of visual observations, and recognized it as a real discrepancy which could not be explained at that time. The later photo-electric observations (made at an effective wave-length of about λ 4500) indicated that the depths of minima found by Wendell were in better agreement with the visual depths inferred from the photo-electric work, and it was pointed out that the conventional comparison of the 1920 visual observations with the 1939 photo-electric observations led to unacceptable conclusions concerning the relative surface brightnesses. Yet the most critical examination of these three sets of observations gives no suggestion that observational error can be responsible for the discrepancy. The accidental scatter is much too small for such an explanation, and systematic errors (such as the Purkinje effect) would affect the depth of primary minimum much more than that of secondary.

Table 2. *Corrected Photometric Elements of R Canis Majoris*

	Wendell (adjusted and rectified)	Dugan (adjusted)	Dugan (corrected for shell)	Wood
$1 - \lambda_1$	0·053	0·038	0·046	0·035
$1 - \lambda_2$	0·412	0·363	0·435	0·405
L_b	0·892	—	0·906	0·928
L_f	0·108	—	0·094	0·072

If we assume, however, that the ejected material was incandescent (from analogy with novae, Wolf-Rayet stars, P Cygni stars and similar objects), quite a different interpretation is possible. We are now in a case similar to that which occurs when a third body contributes to the light of the system, and the observed light curve representing the combined light of the three bodies must be corrected to obtain the true depths of minima that would be found if we could observe the two bodies alone. The chief difference in this case is that, instead of observing the light of the ejected material, we must infer it by asking what its brightness must be in order to bring Dugan's observations into agreement with the other sets.

For a proper comparison we must take one other step: Wendell's observations must be corrected for ellipticity. If we do so, and adjust the depths of minima of each of the sets of visual observations by about 0·01 magnitude in the direction necessary to produce agreement—an adjustment easily permitted by the degree of precision of the observations—and assume that the shell contributes 0·2 of the total light of the system at the time of Dugan's observations, then we find the results shown in Table 2. Reasonable agreement is obtained between the visual observations, and a difference between the visual and photo-electric results is in the direction to be expected from the relative wave-lengths. The entire story may not be quite this simple; nevertheless, the concept of a developing unstable condition, followed by either continuous or intermittent ejection of material for several years does promise at least a partial solution of old and hitherto unresolved discrepancies. It is of interest to note that, in order to rationalize his observations

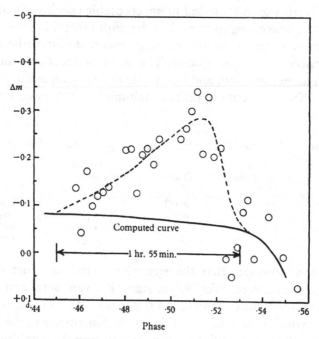

Fig. 1. Photographic observations of SV Camelopardalis on JD 2429287. (In *Princeton Contr.* No. 21, note that the values of $a-v$ at times 0·7480 and 0·7503 should read 0^m·13 and 0^m·14, respectively.)

of W Ursae Majoris, Kwee has found it necessary to postulate a contribution from some other source amounting to 0·25 of the total light of the system.

We ask whether other systems have shown similar photometric humps, and if so, what is their history. A similar effect, but in this case at a phase just preceding primary minimum, has been found in SV Camelopardalis[10], and is illustrated in Fig. 1. The plotted points represent photographic observations taken on one night in 1939. The plates were deliberately taken out-of-focus and the image densities were measured photo-electrically. Each plate was calibrated by impressing on it a series of standard squares whose densities were also measured photo-electrically.

148

A check star was measured at least once on each plate; at the time of the hump, the comparison-check star differences showed only the usual observational scatter. There are available scattered observations on one other night (about a week earlier) at this part of the light curve. These observations also indicate the existence of a hump of about the same amplitude, but suggest that it was then located somewhat closer to primary minimum.

The hump observed in SV Camelopardalis is in many ways similar to that of R Canis Majoris. It was not clearly present on earlier series of observations such as those of Detre [11] and Pierce [12]. In later years, various photo-electric observers covered the curve without reporting any trace of it. Yet, gradually, the later observers began to uncover evidence of other irregularities. Nelson [13] noted asymmetry in the light curve. A Lick Observatory report [14] stated that photo-electric observations indicated evidence of spots similar to those described for AR Lacertae. Hiltner [15] found no hump, but did observe the secondary minimum to be significantly shallower than was found photographically to be the case in 1939, and he also reported irregular surface brightness variations. One spectrogram—but only one—showed weak Ca II K emission. Finally, in an as yet unpublished investigation based on extensive photo-electric observations, H. van Woerden finds variations in the depth of secondary minimum by as much as 0·15 magnitude and also variations in the depth of primary.* Van Woerden also finds evidence of plateaux on the ascending branch of secondary minimum which change in length and slope within a few days. He has investigated thoroughly the older observations, and finds in them evidence of irregularities that had previously escaped attention. He concludes that the light curve has probably displayed irregularities for a long time without our being aware of it. Again, this is a system which has shown erratic period fluctuations. After almost 20,000 epochs of constant period, a sudden change of period occurred, followed by another interval of period constancy, and then by another apparently sudden change at or shortly after the time the hump was observed. More recent photo-electric observations by van Woerden have indicated even larger changes. While the system seems more erratic than R Canis Majoris, the same general picture appears of a temporary hump, period changes, and peculiarities in the light curve.

I know of no other system showing precisely this type of instability. If we look at UX Monocerotis, however, we see evidence of instability on a considerably greater scale. Actually we do not really know whether this

* I am indebted to Dr van Woerden for sending this information in advance of publication.

is a system that is intrinsically more 'unstable', or whether we have been fortunate in catching with all the power of modern observational astronomy a representative system at a particularly unstable phase in its development. The history of this system goes back to 1926. Even in the earliest work based on photographic estimates[16], the dispersion of the observations suggested to the authors that maximum light was not constant. Wyse[17]

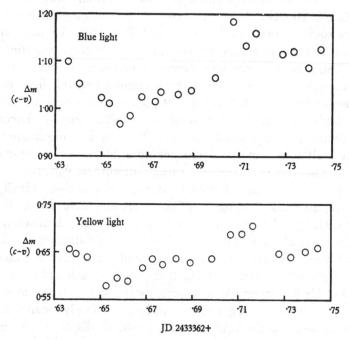

Fig. 2. The light variations of UX Monocerotis in two colours.

classified the spectra as A5 and dG1p and pointed out that the fainter component apparently had a dwarf spectrum and the density of a giant star. Further spectrographic peculiarities, first pointed out by Gaposchkin[18], have been discussed in detail by Struve[19]. The spectral changes are extremely complicated and involve changes in the strength of both the emission and the absorption lines. The irregular light variations were detected quite independently by photo-electric observations at the Steward Observatory[20] and at the McDonald Observatory[21] in the spring of 1950. While the out-of-eclipse light is often relatively stable, it sometimes exhibits surprisingly large fluctuations.

A typical fluctuation is shown in Fig. 2. Over about $2\frac{1}{2}$ hours, variations of almost 0·2 magnitude in the blue and about half this much in the yellow were observed.* On some nights, even more rapid changes were indicated, as illustrated in Fig. 3. On that particular night, the system was observed both at the McDonald and the Steward Observatories. The McDonald observations, corrected to the comparison star used at Tucson, are included. Unfortunately, the McDonald observations were interrupted at precisely the times when the rapid changes were occurring, so the strong confirmation which might be hoped for from simultaneous observation with two telescopes is lacking.

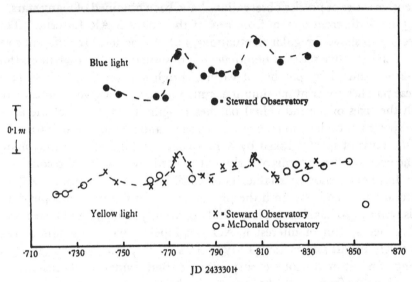

Fig. 3. Observations of UX Monocerotis made at the McDonald and Steward Observatories on the same night.

Recently, C. R. Lynds has observed this star in three colours at the Lick Observatory: his findings strongly confirm this general type of irregular fluctuations. Lynds and B. S. Whitney[22] have independently detected period variations; Whitney has suggested that, since Struve's spectrographic study was made when the period was increasing, it might be valuable to carry out a similar study at a time of decreasing period.

An example even more extreme than UX Monocerotis is shown by AE Aquarii. There is as yet no definite evidence that this is an eclipsing

---

* Publication of these observations has been delayed in the hope of completing the entire curve. This hope has been temporarily abandoned, and I plan to publish the observations in the near future.

variable but Joy's spectrographic work indicates that it is a close double system. The erratic light fluctuations, studied chiefly by Lenouvel[23], are similar in general form to those just described, but are more rapid and much greater in amplitude. In two or three minutes the system has been observed to double in brightness in the spectral range in which it was observed.

Whether there exists a continuous sequence from the stars showing humps on rare occasions (R Canis Majoris) to those exhibiting almost continuous variation (AE Aquarii) and whether, if such a sequence exists, it has evolutionary significance are matters that can be decided only after extensive series of precise observations have been obtained of many stars.

Quite a different system from any of the above is AR Lacertae. This system also shows irregular fluctuations, but of a noticeably different sort. From an extensive series of photo-electric observations, Kron[24] infers that these are caused by patches of varying brightness on the star's surface. These patches cover at maximum as much as 20 % of a given hemisphere, with the sizes of the individual patches ranging from 3–5 % of the area. The spectra of both components are peculiar, and an analysis by Struve[25] on the basis of spectra taken by Sanford and by himself interprets many of the peculiarities in terms of turbulent spottedness. Struve also concludes that the phenomenon is variable. Thus a strikingly similar picture is built up quite independently by both the photometric and the spectroscopic data. This system also has shown a large and apparently sudden period variation.

Another system of interest is AO Cassiopeiae, which consists of two extremely bright and massive O-type stars. The system is obviously quite young, and it may not be surprising to find evidence of instability. Variations are found in the light curve, in the velocity curve, and especially in period. The changes prior to 1947 have been summarized[26]. In 1946–7 the system showed an asymmetric light curve. Observations made in 1947–8, however, indicated nearly equal maxima. The general picture, as yet not strongly established, seems to be that of a normally symmetric curve which is often significantly disturbed. The distortion apparently remains more or less constant during an observing season; at least, shorter term variations have not been firmly established. In 1948 Hiltner[27] observed the star in two colours. His chief conclusions were that the maxima were then nearly equal, the secondary was apparently displaced, and that the two minima were of markedly different shapes. The significant spectral changes of this system have already been discussed[28]. They include both changes in shape of the velocity curve and variation in line intensities; the latter variation was present also in spectroscopic observations made in 1916–17. Contrary to some of the other systems discussed, AO Cassiopeiae has shown recurrent instability for over thirty years. The period changes are also well established [29]; they are certainly not periodic, and seem to occur at irregular intervals. Instead of being of the order of less than a second, as is usual in systems showing such changes, the magnitude of these period variations is of the order of 15 sec. or more. Certainly, whatever is happening, the system is now in process of evolution in an unstable manner.

Turning from these systems which can easily be called unstable and each of which must be described separately, let us look at the close dwarf systems generally classified as of the W Ursae Majoris type. Here, intrinsic variability is the rule rather than the exception. Indeed, I know of no case where precise photometric observations at two or more different epochs have failed to indicate changes in the light curve. Two general types of such fluctuations, not necessarily mutually exclusive, seem to exist.

The first type of change apparently is characteristic of all the W Ursae Majoris variables. A most thorough study has recently been made by L. Binnendijk [30] of 44 Bootis B, a star that may be considered fairly typical of the class. The fluctuation of the relative heights of the quarter-points and the various degrees of asymmetry introduced thereby are clearly evident. It is possible that the displaced secondary minima reported in these systems are really caused by asymmetric light curves that can produce an apparent displacement of secondary.

Two rarer types of fluctuation have been described by Kron [31] for YY Geminorum (the only known dMe eclipsing binary) and by various authors for UX Ursae Majoris [32, 33, 34]. Preliminary reports on DQ Herculis [35] indicate that it may be similar to UX Ursae Majoris. In the case of YY Geminorum, Kron has explained the secondary light variation by a combination of rotation and the eclipse of a patchy, non-uniform surface. In the case of the last two systems, plateaux were found in the light curve which were not constant from night to night and were not explicable by any normal eclipse hypothesis. These systems have been described thoroughly in the recent literature.

At the other extreme of size, systems which have late-type super-giant components commonly show signs of unstability. The study of these has been a fruitful field in recent years, both photometrically and spectro-graphically, and variations other than those caused by eclipse are often truly remarkable. They range from large fluctuations, which in the case of VV Cephei and 32 Cygni may be as large as 0·3 to 0·5 magnitude (and which, except for a longer time scale, somewhat resemble the fluctuations of

UX Monocerotis), through the smaller changes in ζ Aurigae and ε Aurigae (which resemble slow surges of from a few hundredths to a tenth of a magnitude in range) to the constancy shown so far by 31 Cygni. The paucity of our knowledge of these systems is illustrated by the fact that we do not really know as yet whether the character and magnitude of the fluctuations is really dependent on the system itself or upon the particular epoch of

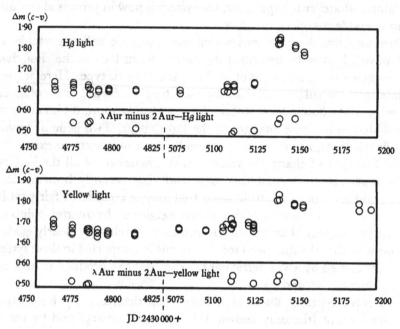

Fig. 4. Observations of ε Aurigae made in 1954 and 1955. The effective wave-lengths of the interference filters used are, approximately, λ 5240 for the yellow magnitudes and λ 4860 for the Hβ observations. The comparison star was λ Aurigae.

observation. In other words, is 31 Cygni, so like others in many of its features but showing no intrinsic light fluctuations, really a stable system or do all these systems show varying degrees of unstability, 31 Cygni having merely been observed at a particularly stable stage? Certainly two solar observers would present somewhat different pictures of the Sun if they observed five years apart.

An example of this type of fluctuation is shown by ε Aurigae. Observations by Huffer[36] preceding the last (1928–30) eclipse showed a hump of amplitude about 0·2 magnitude. Four-colour observations taken at the Cook Observatory of the University of Pennsylvania just before the present (1955–7) eclipse show quite a similar effect. The hump appears in all four colours (Figs. 4 and 5). Its amplitude is greatest in a colour region about

120 angstroms wide that is centred at the Hβ line. One difference between Huffer's observations and the modern results lies in the fact that Huffer reported almost continuous fluctuations while, except for the region of the hump, the Cook observations repeat well from night to night. This is an indication again that observations over many epochs will be required

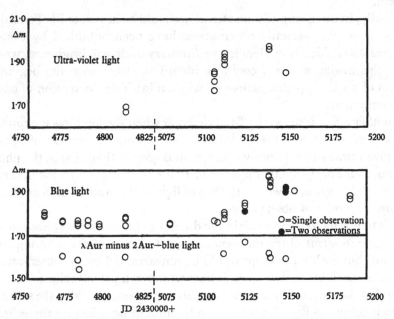

Fig. 5. Observations of ε Aurigae made in 1954 and 1955. The effective wave-length of the glass filter-cell-telescope combination used for the ultra-violet observations is approximately λ 3750. The effective wave-length of the filter used for the observations in blue light is approximately λ 4220. The comparison star was λ Aurigae.

before we can properly describe this system. (Gyldenkerne, at Copenhagen, has also reported irregular variations of this type.)

In addition to these irregular fluctuations, there are other features of these systems that may be connected with symptoms of instability. In an earlier paper [37], Roach and I showed from photometric observations of ζ Aurigae that the duration of the partial phases of the eclipse was a function of the wave-length of observation (as had been suggested by Fracastoro [38] from spectrophotometric results). Thus it was possible to determine the duration of the partial phases only when all the observations had been reduced to a common wave-length. When this was done, it became evident that the width of eclipse was different in 1947 from its value in 1939. (Welsh [39] has summarized the spectrographic observations that establish variations in the extent of the tenuous outer layers.) The change can most simply be explained by a change in size of the K-type component of about 1 %. ζ Aurigae has been intensively observed, yet we do not know if this type of fluctuation is irregular, or part of a periodic change, or even a secular change actually connected with the evolution of the system.

Finally, in this group of stars attention should be drawn to BL Telescopii. Two-colour photo-electric observations have been published by Cousins and Feast [40]. Clearly evident is a preliminary decline in brightness, which, from this evidence alone, could be identified either with the beginning phases of an atmospheric eclipse or with an intrinsic fluctuation of one of the components.

Another effect is shown by BL Telescopii when the light loss is expressed as fraction of eclipse. It is obvious that during ingress, the loss of light at any given time in the photo-visual region is greater than that in the photographic: an effect exactly opposite to that of ζ Aurigae. In the observed phases of the egress, however, the loss of light is the same in both colours to within the errors of observation.

It is tempting to try to explain the difference between BL Telescopii and ζ Aurigae in terms of the difference in spectral class of the eclipsing components, but such an attempt would be unwarranted by the observational evidence available to date. BL Telescopii shows a partial eclipse. Instead of forcing the two curves together at mid-eclipse, one should make solutions in each colour to find the maximum fraction of light lost by the eclipsed component, and compute the 'fractions of eclipse' from this. Two difficulties arise, however. The first is that solutions by conventional methods may not be appropriate in this case. The second difficulty is that the South African observers have discovered intrinsic fluctuations in the light of BL Telescopii which make it difficult to determine the uneclipsed brightness of the system. But any consistent treatment leads to the same qualitative conclusion, that there is an observable difference between ingress and egress at the same phases. This system warrants intensive investigation.

Time does not permit a complete discussion of other systems that have shown indications of non-stable features. In BM Cassiopeiae, for example, two-colour photo-electric observations by G. Thiessen have confirmed earlier reports by M. Beyer. Fluctuations in depth and shape of the minima and considerable fluctuations outside eclipse are found.

Many other unusual systems exist. One has only to name RT Andromedae, SX Cassiopeiae, RX Cassiopeiae, RT Coronae Borealis, W Serpentis, GO Cygni, V444 Cygni, V729 Cygni, V367 Cygni, VW Cephei, and

U Pegasi to demonstrate that already we know many systems that have shown unstable characteristics and which will repay further observation.

Let us turn finally to a feature found in many eclipsing systems: erratic period changes that do not fit any concept of a stable system. It seems to be significant that in almost every case where well-established irregular period changes exist in systems for which photometric solutions are available, one component of the system is near the limits of dynamical stability. The material available through 1950 has already been sum-marized [9]. I want to report my strong conviction that the study of these period changes may be of great significance in attacking the difficult pro-blem of the evolution of close binary systems.

In summary, then, today we seem to be in the initial stages of exciting new developments in the study of eclipsing binaries. Two years ago, I pointed out that the study of eclipsing systems could roughly be divided into three phases. In the first, the theory was weak or non-existent and the observers frequently reported wonderful features in the light curves: features that we now know were usually characteristics of the observing technique rather than of the star. This phase ended with the theoretical papers of Russell in 1912–13, and, observationally, with the development of visual photometers, more accurate photographic techniques, and the use of photo-electric cells. For more than thirty years, we were in a second period in which the effort of the theoretical worker was directed toward ex-planation of the light fluctuations by more and more highly refined models, and the observer tried to produce curves of the highest possible precision, usually serene in the belief that the light variations of the star were repeating themselves uniformly cycle after cycle and year after year. About 1946 the stimulating spectrographic work of Struve and his collaborators on the one hand, and several series of photo-electric observations on the other, called attention strikingly to the fact that changes are occurring in these systems. We are now aware that few of the eclipsing systems now known are far removed from the limits of dynamical stability (although we should remember that observational selection gives preference to the discovery of such systems), and that we may have a chance to study evolutionary changes without waiting for millions of years.

In addition to using these ideas in planning our own programmes and interpreting the results, it may be of value to re-discuss the observations of the past in order to see what information this new approach may produce. We will of necessity be limited to cases where the observational technique was such as to minimize the systematic errors and where the individual observations were themselves published. In general, these will be systems observed photo-electrically, with the visual polarizing photometer or in some cases with the wedge photometer, or photographically when a sound objective technique was used in measuring the magnitude changes.

The importance of publishing the individual observations can scarcely be over-emphasized. Only thus will observations made now be available with their full potential to astronomers of the future.

REFERENCES

[1]   E. C. Pickering, *Harv. Ann.* **46**, 172 (1904).
[2]   O. C. Wendell, *Harv. Ann.* **69**, 66 (1909).
[3]   R. S. Dugan, *Princeton Contr.* No. 6, 59 (1924).
[4]   F. C. Jordan, *Allegheny Publ.* **3**, 49 (1916).
[5]   R. H. Hardie, *Ap.J.* **112**, 542 (1950).
[6]   B. W. Sitterly, *A.J.* **48**, 190 (1940).
[7]   O. Struve and B. Smith, *Ap.J.* **111**, 27 (1950).
[8]   F. B. Wood, *Ap.J.* **112**, 196 (1950).
[9]   F. B. Wood, *Princeton Contr.* No. 21, 31 (1946).
[10] F. B. Wood, *Princeton Contr.* No. 21, 46 (1946).
[11] L. Detre, *A.N.* **249**, 213 (1933).
[12] N. L. Pierce, *Princeton Contr.* No. 18, 3 (1938).
[13] B. Nelson, *A.J.* **56**, 136 (1951).
[14] *Publ. A.S.P.* **62**, 40 (1950).
[15] W. A. Hiltner, *Ap.J.* **118**, 262 (1953).
[16] I. E. Woods and M. B. Shapley, *Harv. Bull.* No. 854, 6 (1928).
[17] A. B. Wyse, *Lick Obs. Bull.* **17** (No. 464), 37 (1934).
[18] S. Gaposchkin, *Ap.J.* **105**, 258 (1947).
[19] O. Struve, *Ap.J.* **106**, 255 (1947).
[20] F. B. Wood, *A.J.* **55**, 187 (1950).
[21] W. A. Hiltner, O. Struve and P. D. Jose, *Ap.J.* **112**, 504 (1950).
[22] B. S. Whitney, *Publ. A.S.P.* **68**, 253 (1956).
[23] F. Lenouvel and J. Daguillon, *Ann. d'Ap.* **17**, 416 (1954).
[24] G. E. Kron, *Publ. A.S.P.* **53**, 261 (1947).
[25] O. Struve, *Publ. A.S.P.* **64**, 20 (1952).
[26] F. B. Wood, *Ap.J.* **108**, 28 (1948).
[27] W. A. Hiltner, *Ap.J.* **110**, 443 (1949).
[28] O. Struve and H. G. Horak, *Ap.J.* **110**, 447 (1949).
[29] F. B. Wood, *A.J.* **56**, 53 (1951).
[30] L. Binnendijk, *A.J.* **60**, 355 (1955).
[31] G. E. Kron, *Ap.J.* **115**, 301 (1952).
[32] A. P. Linnell, *Harv. Circ.* No. 455 (1950).
[33] M. F. Walker and G. H. Herbig, *Ap.J.* **120**, 278 (1954).
[34] H. L. Johnson, B. Perkins and W. A. Hiltner, *Ap.J. Suppl.* **1** (No. 4), 91 (1954).
[35] M. F. Walker, *Publ. A.S.P.* **66**, 230 (1954).
[36] C. M. Huffer, *Ap.J.* **76**, 1 (1932).
[37] F. E. Roach and F. B. Wood, *Ann. d'Ap.* **15**, 21 (1952).
[38] M. G. Fracastoro, *Arcetri Oss. e Mem.* No. 63, 83 (1945); *Harv. Circ.* No. 456 (1951).
[39] H. L. Welsh, *J. Roy. Astr. Soc. Canada*, **43**, 217 (1950).
[40] A. W. J. Cousins and M. W. Feast, *Observatory*, **74**, 88 (1954).

20. NON-STABILITY IN CLOSE BINARY STARS

V. A. KRAT

Pulkovo Observatory, Leningrad, U.S.S.R.

Those eclipsing binaries whose components are non-stable stars deserve particular attention not only owing to the fact that in such cases the physical properties of the stars may be studied in greater detail, but also since the mechanism of the ejection of gases from the atmospheres of the components may then be established with some certainty.

According to our cosmogonical views[1], stars forming a close pair must have originated simultaneously because if their origin from diffuse matter did not occur at the same time, then the second component, while having not yet acquired the properties of a real star, would be torn to pieces by the tidal forces of the main star and dissipate under the influence of its emission. It should be noted that the 'black sphere' temperature of the region in which the second star is formed will be about two times less than the surface temperature of the main star.

The stage of non-stability plays the part of a jump in the evolution of a star. In the course of cosmogonically short periods of time of the order of 10^5 or 10^6 years, a non-stable star undergoes larger changes than those that would take place in ordinary stable stars in the course of 10^8 or 10^9 years[2].

Ejection of gases from the atmospheres of the components of eclipsing variables was discovered by O. Struve. The phenomenon of non-stability in eclipsing variables was considered until recently as being something extraordinary; ejection of gases had been found only in the cases of a few white super-giants(β Lyr) and cooler giants (RX Cas and SX Cas). At present no systems are known to us that have two stable hot super-giants as components, except Y Cyg. The non-stability of white super-giants was recently studied by A. N. Dadaev[3] in the systems of AO Cas and β Lyr. According to N. M. Goldberg-Rogozinskaya[4], u Her, an eclipsing star with a rather regular light curve, is also non-stable. The non-stability of the secondary star in the system of U Cep, discovered by O. Struve, and that of the secondary (a typical sub-giant) in V505 Sgr, discovered by A. V. Sofronizky[5], is of utmost importance. This fact testifies that not only giant stars may be unstable.

Still more interesting is a study by K. Kaltchaev, who investigated systems in which both components are sub-dwarfs: UX UMa, AK Her, AG Vir and RW CrB. Kaltchaev established with certainty that these stars are real sub-dwarfs. This confirmed our former statements that sub-dwarfs are never found in binaries together with stars belonging to other sequences of the H–R diagram and thus of a different age. All sub-dwarf systems studied by Kaltchaev were found to be unstable. A very important feature of the instability of close pairs is the fact that the non-stability is caused by the circumstance that the surface of one star is very near to the internal critical Roche surface. In all cases in which instability has been established, the photosphere of one of the components touches Roche's internal surface. The system of UX UMa, for which the spectroscopic observations suggest the presence of an extended atmosphere around one of the stars, is the only exception. Spectroscopic observations show that the velocity of ejection is very great in all the systems investigated, being of the order of 100 km./sec.

The rapid decrease of the mass of a non-stable star indicates that the process of ejection began comparatively recently (10^5 or 10^6 years ago) and was preceded by an expansion of the star up to the dimensions of Roche's critical configuration. Thus, the instability of binary systems appears to be a proof of the expansion of the stellar envelopes in the process of their evolution. Such stars could initially be stable stars in which the equilibrium of the outer layers was afterwards disturbed. We believe that white super-giants pass a certain stage in their evolution in which they are stable stars. Thus, for example, the O-type systems of Y Cyg and AO Cas are extremely similar in mass and spectral type; they might be regarded as being in different stages of their evolution, one of them being stable (Y Cyg) and the other non-stable (AO Cas). It is observed, as a rule, that only one star of a binary system is non-stable; the other star does not show any signs of non-stability.

As we indicated earlier[1], white super-giants are non-stable stars which in the course of their evolution enter into the stage of unsteady stars (probably Wolf-Rayet stars). The instability of these stars is caused by the lack of equilibrium of their outer layers through the so-called corpuscular instability. By corpuscular instability is understood a state of the stellar atmosphere in which the atoms may dissipate at thermal velocities. Corpuscular instability sets in either as a consequence of a low gravity in the atmosphere of a star, or in consequence of a high temperature. In the atmospheres of white super-giants both factors are present. Such stars must be corpuscularly unstable, as are the majority of super-giants and red giants. It should be noted that any star that reaches the critical Roche surface during its expansion must automatically become corpuscularly unstable. The ejection starts in this case from the zone of the internal critical point.

Corpuscular instability causes a disturbance of the equilibrium of the gaseous masses near the surface of the star. The cause of such a disturbance is the enormous inertia of the luminosity of a star. The luminosity of a star, owing to the slowness of the energy transfer through the gaseous layers, does not correspond to the activity of its present energy sources. A star suddenly deprived of energy sources will still maintain its luminosity for about 10^6 or 10^7 years. The loss of mass by a star at the surface will therefore lead to no change of the luminosity of the star during an interval of time of less than 10^6 or 10^7 years, even if the activity of the energy sources has changed. The evolution of the star during this interval of time takes place under contradictory conditions of constant luminosity and decreasing mass. It is easy to show that in the case of constant luminosity the process of ejection will be a self-accelerating process. At an initial moment at some equipotential surface $R = R_0$ near the boundary of the star let there exist hydrostatical equilibrium: that is, let the condition

$$\frac{1}{\rho} \operatorname{grad} P = g \tag{1}$$

be fulfilled, where ρ is the density, and P is total pressure. In a first approximation the gas pressure is

$$P = \frac{R}{\mu} \rho T, \tag{2}$$

where μ is the average atomic weight, T is the absolute temperature, and g is the acceleration, representing the sum of the Newtonian acceleration and centrifugal acceleration. Starting from the surface $r = R_0$ and calculating r in the direction towards the centre of the star, we can assume, by neglecting the second component of the acceleration, that

$$g \approx \frac{Gm}{R_0^2}, \tag{3}$$

where m is the mass of the star. During a time interval dt the mass dm will be lost from the atmosphere of the star through dissipation. The left side of the equation (1) will then be changed as follows:

$$d\left(\frac{1}{\rho} \operatorname{grad} P\right) = \frac{R}{\mu} T (d \operatorname{grad} \ln T + d \operatorname{grad} \ln \rho). \tag{4}$$

Since in any model of a stellar atmosphere, $\ln \rho$ is always undergoing a greater change than is $\ln T$ (on the boundary in particular), it is advisable to take in the right side of the equality (4) the member with d grad $\ln \rho$ only. But when ejection of matter takes place, d grad $\rho > 0$ and therefore

$$d\left(\frac{1}{\rho} \text{ grad } P\right) > 0. \tag{5}$$

At the same time, if the mass changes by dm, then $dg < 0$. The left side of the equation (1) will exceed the right one, and the star will begin to expand. It is also obvious that if the process is reversed and the star accumulates mass, then d grad $\ln \rho$ must decrease and g grows, and a disturbance of the equilibrium will lead to a compression of the star.

Owing to the slowness of the change of velocity v on the surface $r < R_0$ with time, the ejection can be considered as a stable process, the parameters of which undergo a gradual change with time. In this case the change of the kinetic energy in the stream equals the change of the potential energy, and the change of the internal energy of the gas can be neglected in a rough approximation. It will be easy then to calculate the increase of the velocity of the stream for a given change of mass. As was shown in reference 1, even for deliberately lowered estimates of the decrease of the mass of corpuscularly unstable stars, the ejection which goes on initially with the thermal velocity of the order of 10 km./sec., will after 10^6 years reach the value of 10^2 or 10^3 km./sec. The star will become a typical Wolf-Rayet star. A permanent decrease of temperature and pressure in the central regions of the star owing to the weakening of the thermonuclear reactions will stop the process of expansion. At this stage the star will be a red super-giant or giant and will lose mass only by dissipation. Probably the process of expansion will be replaced by contraction which might stop only in the case when the star acquires equilibrium. Judging by the fact that in the Wolf-Rayet stage and then in the red giant stage the star might decrease in mass by several times, this new state of equilibrium will correspond to that of a main sequence star [2].

It is interesting to observe that the process of mass accumulation on the surface of the star leads to the same result but to a reverse order of the processes of evolution. First, a contraction of the star will take place. T and P will grow in the centre of the star, causing an increase in the rates of thermo-nuclear reactions. After a short period of time (equalling the time required for the contraction, starting on the surface, to reach the centre), an expansion of the internal and, afterwards, the external layers of the star will occur.

It is also interesting to note that the gas stream flowing from the zone of the internal critical point of the critical Roche configuration is, contrary to

162

Kuiper's view [6], asymmetrical and can surround the second star only at low velocities. In most cases it dissipates soon after its emergence from the zone of the internal critical point. This follows from theoretical deductions: the stream must possess a small angular momentum, corresponding to that in the critical point region. It also follows from observation that a decrease of light in the stream always causes asymmetry in the light curve out of eclipse. The well-known periastron effect, expressed by the differences in the maxima of the light curve and formerly attributed to the change of the phase effect when a star is moving in its elliptical orbit, is now almost always connected with the stream from the components. The new interpretation of the mechanics of the ejection of gas from non-stable stars makes it possible to explain the variation of the period of β Lyr by the loss of mass, and to estimate the age of that system as about 10^6 years [2].

In conclusion, we shall attempt to give a new cosmogonical interpretation of the H–R diagram based mainly upon data obtained from studies of eclipsing variables.

The evolution of massive stars proceeds from the stage of stable hot super-giants through the stage of Wolf-Rayet stars and red giants to the main sequence (thus approaching solar-type stars). The mass decreases according to the law

$$\frac{d\mathbf{m}}{dt} = 4\pi R^2 \rho_0 v_0, \qquad (6)$$

where v_0 is the rate of ejection, ρ_0 is the density at the surface of the star, and R is the radius of the star. As soon as v_0 and R begin to grow progressively in the Wolf-Rayet stage, the derivative $d\mathbf{m}/dt$ reaches a maximum. In the giant stage, it decreases rapidly and reaches zero in the stage of solar-type stars. We do not believe an evolution along the main sequence to be possible. E. R. Mustel showed recently [7] that solar-type stars lose mass at a rate of the order of 10^{16} grams/year. Stars of small mass in their initial stage of development are not met among the eclipsing variables known to us. Perhaps only the system of YY Gem, in which the spectra of the components have emission lines typical of the T Tau variables (which, according to Ambartsumian, form T-associations and are therefore very young stars), may be thought of as young non-stable stars of small mass. The process of expansion of the initially stable star transforms it into a sub-giant. The loss of mass by the star might reach its largest value in that stage of the star's evolution. For stars of small mass, sub-giants may play in a number of cases the role of Wolf-Rayet stars. The process of contraction makes the sub-giant evolve into a star belonging to the second part of

11-2

the main sequence, according to Parenago; i.e. into a star with comparatively large density and mass less than the solar mass. If the energy source of such stars is the proton-proton reaction, then their further transition into the sub-dwarf stage [1] will be altogether natural. According to Kaltchaev, sub-dwarfs are stars of small mass and radius, and large density by comparison with main-sequence stars (the main characteristics of main-sequence stars were taken from Parenago's data). It is obvious that main-sequence stars transformed into sub-dwarfs were also objects of small mass and a still greater density, because when the amount of helium increases, the opacity of the star increases also and expansion takes place. The fact that all eclipsing variables composed of sub-dwarfs are found to be unstable suggests that the cause of their instability is the passage of a main-sequence star into the sub-dwarf phase, the star expanding and becoming unstable after it has reached Roche's limiting figure. We can, therefore, believe that in this case the instability of the star does not suggest its youth but rather its comparatively old age. Non-stability can arise in various stages of stellar evolution.

REFERENCES

[1] V. A. Krat, *Pulkovo Bull.* **19**, No. 2, 1 (1952).
[2] V. A. Krat, *Pulkovo Bull.* **18**, No. 4, 1 (1950).
[3] A. N. Dadaev, *Pulkovo Bull.* **19**, No. 5, 31 (1954).
[4] N. M. Goldberg-Rogozinskaya, *Pulkovo Bull.* **18**, No. 6, 64 (1951).
[5] A. V. Sofronizky, *Pulkovo Bull.* **19**, No. 4, 1 (1953).
[6] G. P. Kuiper, *Ap.J.* **93**, 133 (1941).
[7] E. R. Mustel, *Publ. Crim. Astrophys. Obs.* **10**, 143 (1953).

21. THE CHROMOSPHERIC SPECTRUM AND THE ATMOSPHERE OF 31 CYGNI

A. McKELLAR AND R. M. PETRIE

Dominion Astrophysical Observatory, Victoria, B.C., Canada

The binary 31 Cygni is not, in the ordinary sense, an unstable star. Nevertheless, the organizers of this symposium have invited the present paper as a discussion of additional recent evidence of random motions of masses of gas in the atmospheres of the late-type giant components of certain double stars.

31 Cygni is an eclipsing binary similar in general nature to the well-known system, ζ Aurigae. It consists of a super-giant K-type primary and a secondary of type B5 or possibly a little earlier, according to Wright and Lee. The two stars are equally bright near λ 3900. While 31 Cygni has long been known as a spectroscopic binary, its eclipsing nature was predicted by D. B. McLaughlin only about five years ago from an examination of spectrograms taken in 1941. The period is about 3800 days. The only well-observed eclipse occurred in 1951 when series of spectrograms were obtained at Victoria, Ann Arbor and Stockholm, and some fragmentary photo-electric observations were made.

Publications on 31 Cygni may be grouped into earlier investigations of the spectrum, radial velocities, and orbits [1-4], the suggestion that the star was an eclipsing binary [5], the photo-electric observations [6], and papers describing results obtained from the Ann Arbor [7, 8], Stockholm [9] and Victoria [10-13] spectrograms of 1951. The material given in the present discussion is taken from the three papers: (*a*) reference [10] above; (*b*) a paper being prepared by R. M. Petrie and A. McKellar on the Victoria observations of 31 Cygni; and (*c*) a paper being prepared by A. McKellar, L. H. Aller, G. J. Odgers, and E. H. Richardson on the chromospheric K-line of Ca II in the spectrum of 31 Cygni. The latter two papers are to appear as *Publications of the Dominion Astrophysical Observatory*, vol. XI, Nos. 1 and 2.

In 1951 spectroscopic effects of the impending eclipse of 31 Cygni were observed as early as 1 June. The chromospheric lines increased in intensity, slowly at first but rapidly in early August during the final days of ingress. Total eclipse of the comparatively small B-type star began on 12 August and lasted until 12 October. During egress, the spectroscopic phenomena of atmospheric eclipse occurred in the reverse order and were detectable until the end of the year. Combination of the existing orbital and photometric data gives the minimum diameter of the K-type super-giant component as 150⊙ and the diameter of the B-type star as about 5⊙. It may be remarked that our preliminary measurements on the secondary spectrum indicate a somewhat larger mass ratio than has been given and so would require larger minimum sizes for the stars.

Most of the spectrograms photographed at Victoria were obtained with the third order of a 15,000-lines-per-inch grating and a collimator-camera lens of 45 inches focal length in a Littrow mounting, giving the comparatively high dispersion of 4·6 Å/mm. The wave-length region covered was from λ 3700 to λ 4600 except for a gap from λ 4100 to λ 4200.

Study and interpretation of the spectrograms has shown, in at least three ways, evidence of structure and motions in the outer atmosphere of the super-giant star. The three will be briefly described. Fuller details may be read in the papers cited above. The term 'chromospheric' as used above and throughout this paper refers to the extra absorptions present in the composite spectrum due to the light from the early-type star traversing the atmosphere of the late-type super-giant for periods before and after total eclipse. While it is not implied that the outer atmosphere of the primary component is similar to the solar chromosphere, the term has seemed apt since, except for scale, certain similarities do exist.

(1) STRUCTURE AND MOTIONS IN THE OUTER CHROMO-SPHERE DEDUCED FROM COMPLEX K LINES

The chromospheric K line of Ca II during egress in late October and in November and December 1951, showed definite structure. From 26–31 October, the principal wide chromospheric line was accompanied by a weaker wing or satellite which, with respect to the rectangular-shaped principal line, occupied positions corresponding to radial-velocity shifts as great as 130 km./sec. Also, at various times from mid-November until late December, the K line was often double, the components varying in intensity and separation from day to day and week to week. The separation in wave-length corresponded to radial-velocity differences of 22–44 km./sec.

These observations have already been described [10] and have been interpreted as evidence of the existence of discrete moving clouds of material in the outer chromosphere of the K-type giant star.

The reproductions in Fig. 1 show spectrograms in the region λ 3878 to λ 3980 for the period 31 July to 10 August 1951, just preceding totality, as well as a spectrogram in totality and one almost a year later.

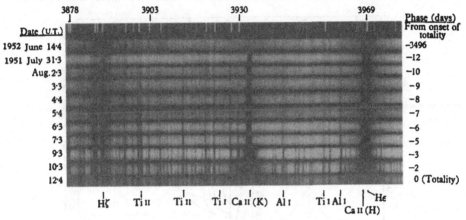

Fig. 1. Chromospheric spectrum λ 3878–λ 3980, late stage of ingress, 1951.

(2) THE 'TURBULENT' VELOCITY INDICATED BY THE PROFILES OF THE CHROMOSPHERIC K LINE

Intensity profiles have been derived for the chromospheric K line from the Victoria plates covering ingress and egress and from the Ann Arbor plates obtained during egress. These line profiles have been compared with calculated profiles.

As shown by examples given in Fig. 2 reasonably good accord between observed and calculated profiles is secured using values of turbulent velocity ranging from 20 km./sec. for the low chromosphere to 10 km./sec. for the masses of material in the high chromosphere. It may be remarked that Miss Underhill[11] found the widths of the four Fe I chromospheric lines she studied, which originated in the lower atmosphere of the K-type star, to correspond to motions up to 20 km./sec.

When the equivalent widths of the K line are plotted against the line depths (corrected for instrumental effects), as shown in Fig. 3, we obtain further evidence of motion among the Ca II atoms in the outer chromosphere of the giant star. On some days the K line shows no structure, and its shape corresponds to a calculated profile for a v_T of about 10 km./sec. However, on other days both before and after eclipse the line is not actually double, but shows some indication of unresolved structure, such as a lack of symmetry. Then the relation between equivalent width and depth, as shown by the triangular symbols grouped toward the right side of Fig. 3,

167

indicates values of turbulent velocity of 20 km./sec. or more. In these cases the larger velocities presumably arise from absorption in two or more masses of gas having different velocity distributions in the line of sight. The line of sight is, of course, tangential to the stellar surface. The relative

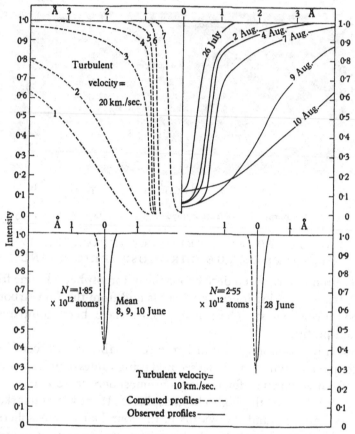

Fig. 2. Observed and calculated profiles of the chromospheric K line. In the upper half of this figure the relation of the profile number to the number of atoms is as follows: 1, 355×10^{16}; 2, 177×10^{16}; 3, 22×10^{16}; 4, $5 \cdot 8 \times 10^{16}$; 5, $1 \cdot 9 \times 10^{16}$; 6, $0 \cdot 18 \times 10^{16}$; 7, $0 \cdot 002 \times 10^{16}$.

velocities of the components are sufficient to widen the line but not to produce definite multiplicity.

(3) RADIAL-VELOCITY MEASUREMENTS ON THE CHROMOSPHERIC LINES

The most recently completed section of the work on the spectrum of 31 Cygni is the measurement, for radial velocity, of the chromospheric lines in the spectrum from twelve days before total eclipse until onset of totality and from the end of totality for a period of fourteen days. The coverage during ingress is fairly complete, but during egress it is fragmentary.

The plates measured were all third-order grating plates (4·6 Å/mm.) except those of 16 October, for which date three prismatic spectrograms of

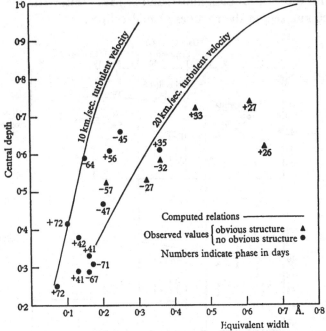

Fig. 3. Relationship between the central depth and the equivalent width of the chromospheric K line.

lower dispersion, 20 Å/mm. at λ 3933, were available. For the grating plates, measurements on the spectrum of α Persei indicate the probable error of the velocity from a single line of average weight to be ± 3 km./sec.

Only lines clearly chromospheric in character were measured. Lines known or suspected to be blended were omitted. The Balmer lines of hydrogen were surprisingly badly affected by blending. The numbers and origin of chromospheric lines measured (those which survived the inspection for blending and chromospheric character) are shown in the following tabulation:

Atom	Fe I	Ti I	Ti II	Mn I	Cr I	Ni I	Al I	Mg I	Ca II	Sr II	Sc II	V I	Y I	Ba II	Ca I
No. of Lines	46	10	7	3	4	3	2	3	2	2	2	1	1	1	1

The results of the measurements are shown graphically in Fig. 4. The time scale used for the period of total eclipse is only one-third that used in the rest of the figure. The most reliable results, those of Fe I, Ti I and Ti II, are shown at the top, the next most reliable group is in the middle, and the least reliable at the bottom of the figure. For each group the segment of the radial-velocity curve of the orbital motion of the K-type star is given. This radial-velocity curve, as redetermined at Victoria, agrees exactly in slope with those of McLaughlin [8] and of Miss Vinter Hansen [4], and is within 0·26 km./sec. of their values at mid-eclipse.

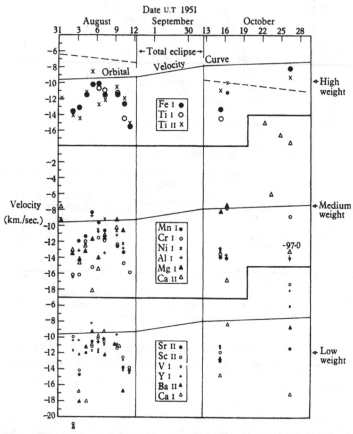

Fig. 4. The radial velocity of 31 Cygni in the vicinity of eclipse from chromospheric lines.

It is apparent that the radial velocities measured for the chromospheric lines fall systematically below the orbital curve, both before and after eclipse, by several km./sec. The effect as shown by the most reliable group is surprisingly well duplicated by the other two groups. Since the orbital velocity curve was determined from similar spectrograms during totality and at phases far removed from totality and since, also, the grating spectrograms gave the accepted velocity for the standard velocity star α Persei, this discrepancy is considered to be an actual velocity effect. Furthermore, in their general behaviour, our measurements agree with those of McLaughlin[8] over the same time interval.

The deviations of the velocities from the orbital curve do not correspond to simple stellar rotation. For rotation, the deviations would be of opposite sign before and after totality, which is not the case. For rotation in the same sense as orbital motion, the deviation at ingress would be positive, not negative as observed. In the top section of Fig. 4, the broken lines show the minimum rotational-velocity effect, assuming equal rotational and orbital periods and similar directions of motion.

We are inclined to attribute the deviations in velocity to tangential components of atmospheric motions, possibly random or possibly partly systematic. The source or driving force of the motions cannot at present be specified, but since the gravitational acceleration at the surface of the super-giant star is unlikely to exceed 10 cm./sec.^{2}, and the atmosphere has a very low mean density, irregular horizontal drift motions could be easily initiated and would not be quickly damped out. Tidal forces are much too small to be an important factor in the problem. Velocity differences between chromospheric and stellar lines, similar to those just described for 31 Cygni, have long been known for ζ Aurigae. It will be interesting and important to examine whether deviations in velocity of chromospheric lines near the 1961–2 eclipse of 31 Cygni will be similar in sign and magnitude to those found for the 1951 eclipse. If so, some systematic source can be sought; if not, the motions are presumably random.

We are left with the following picture of the atmosphere of the super-giant K-type star. The inner chromosphere observed for about twelve days before or after totality, and extending approximately one-fifth of a diameter above the main body of the star, shows evidence of motions with velocities of the order of 20 km./sec. from K line profiles, and of components of motion up to several km./sec., given by radial-velocity deviations of chromospheric lines from the orbital radial-velocity curve. Evidence of occasional high-speed outbursts of gas from the lower chromosphere is given by the satellite lines of the chromospheric K line (26 October to November 1951), that show displacements corresponding to as much as 130 km./sec. The measured motion is the component tangential to the stellar surface. Finally, in the outer atmosphere there must be discrete clouds of gaseous material having, on present evidence of double chromospheric K lines, relative velocities up to 40 km./sec. in the line of sight.

REFERENCES

[1] A. C. Maury, *Harv. Ann.* **28**, 93 (1898).

[2] W. W. Campbell, *Lick Obs. Bull.* **1** (No. 4), 22 (1901).

[3] W. H. Christie, *Ap.J.* **83**, 433 (1936).

[4] J. M. Vinter Hansen, *Ap.J.* **100**, 8 (1944).

[5] D. B. McLaughlin, *Publ. A.S.P.* **62**, 13 (1950).

[6] F. B. Wood, *A.J.* **58**, 51 (1953).

[7] D. B. McLaughlin, *Publ. A.S.P.* **64**, 109, 173 (1952).

[8] D. B. McLaughlin, *Ap.J.* **116**, 546 (1952).

[9] G. Larsson-Leander, *Stockholm Obs. Ann.* **17**, No. 5 (1953).

[10] A. McKellar, G. J. Odgers, L. H. Aller and D. B. McLaughlin, *Nature,* **169**, 990 (1952); *Contr. Dom. Astrophys. Obs.* No. 24.

[11] A. B. Underhill, *M.N.* **114**, 558 (1954); *Contr. Dom. Astrophys. Obs.* No. 37.

[12] A. McKellar, *Proceedings Nat. Sci. Foundation Conference on Stellar Atmospheres at Indiana University*, p. 169, Bloomington, 1954.

[13] K. O. Wright and E. K. Lee, *Publ. A.S.P.* **68**, 17 (1956); *Contr. Dom. Astrophys. Obs.* No. 45.

[14] See, for example, A. McKellar and R. M. Petrie, *M.N.* **114**, 641 (1952); *Contr. Dom. Astrophys. Obs.* No. 29, where references to earlier work on the spectrum of ζ Aurigae are given.

DISCUSSION

Following the papers on instability in binary systems, Dr S. Gaposchkin described a classification scheme that he had devised to describe the eclipsing binaries having emission lines, gas streams, or thick atmospheres. Such stars Gaposchkin calls the 'camouflaged eclipsing variables', and divides them into five groups:

(1) Typical members are UX Mon and SX Cas: the spectral type of the brighter star is B or A, that of the fainter is G or later. The minima are of equal depth. Balmer emission is present.

(2) Prototypes are β Lyr and v Sgr: only one component can be seen distinctly, but two minima are present in the light curve. The spectroscopic behaviour is most complicated.

(3) Prototypes are RT And and YY Gem: the spectral class is G or later. Ca II emission is present. The depths of the minima are roughly equal, and the dimensions of the components are about the same.

(4) The prototype is DN Ori: both components are so well camouflaged that no definite velocity variation has been observed.

(5) Prototypes are V444 Cyg or UX UMa: the bright lines are strong, and their displacements and intensities do not vary in a simple manner with phase.

Dr Z. Kopal, in commenting upon the communication by Dr Wood, emphasized a need for caution in attributing every photometric irregularity of an eclipsing system to the effects of dynamical instability. Kopal felt that, in particular, considerable reserve should still be exercised in associating the secular loss of mass, which probably occurs at the point of instability of an expanding sub-giant, with the changes of period as are often exhibited by such systems. Numerical integrations of the trajectories of such gas streams reveal that the principal effect of these mass motions is to transfer material gradually from one component to the other. The formation of an outer ring of gas encircling the whole system is another possibility. Kopal stated that it is only under very extreme (and unlikely) conditions that any mass can be lost to the system as a whole, and it is only such a loss that could produce an observable lengthening of the period.

Kopal said that under very general conditions the surfaces of zero velocity remain closed about the system, and no secular loss of mass by the system as a whole can be expected as a result of gravitational forces.

Kopal concluded that if the orbital periods of eclipsing systems fluctuate (often varying in jumps rather than continuously), then the real cause of the perturbations of their semi-major axes must be sought elsewhere. Similarly, if ejection of matter with velocities exceeding that of escape is established spectroscopically (photometric observations alone cannot decide this point), an understanding of that phenomenon would have to await consideration of other effects.

Dr Struve, in response to Kopal's remarks, stated that if anything is certain in the spectroscopic study of eclipsing binaries, it is that gas streams do exist in these systems! He also called attention to the fact, as a matter of historical record, that it was Wood, in a 1950 paper, who pointed out the possibility that material might be lost to eclipsing systems through Jacobian surfaces.

Kopal, in reply, said that he did not intend to question the reality of gas streams, so beautifully demonstrated by Struve's observations. He hoped that Struve would agree, however, that the significance of these streams for photometric observations in *continuous* radiation (as opposed to discrete frequencies) is still largely hypothetical. Kopal felt, in particular, that gas streams capable of producing continuous emission that would amount to an appreciable fraction of the total light of an eclipsing system would appear to be of inordinately large mass.

Kopal stated also that, in the interests of historical accuracy, we should perhaps recall that the first investigator who realized the relevance of the Jacobian surfaces of zero velocity for an eclipsing system appears to have been K. Walter, in the case of R Canis Majoris. The authority for this statement was H. N. Russell, in his Harvard Centennial Symposium lecture of 1946.

Dr W. A. Hiltner called attention to the fact that the loss of mass by binaries can actually be observed in the case of the Wolf-Rayet systems. Kopal agreed, but pointed out that it is in precisely these binaries, in which a loss of mass is best established, that no period increases at all have been observed.

VI. PHYSICAL SOURCES CAUSING INSTABILITY

22. STARS OF T TAURI AND UV CETI TYPES AND THE PHENOMENON OF CONTINUOUS EMISSION

V. A. AMBARTSUMIAN
Burakan Observatory, Erevan, Armenia, U.S.S.R.

Our knowledge concerning stars of the T Tauri and UV Ceti types has been considerably enlarged during recent years, owing to the accumulation of a large number of facts. As a result, we may now make certain suggestions regarding the causes of the non-stability of these stars, based on observational data. Such an approach to the solution of the problem is altogether different from that applied heretofore in attempts to develop a theory of certain types of variable stars. The method of construction of models has usually been applied in the past. We do not exclude, of course, the possibility of solving a problem on the basis of some theoretical model. However, the physical phenomena observed in the T Tauri and UV Ceti stars are so unusual that it is first of all necessary to define the nature of these phenomena and their mutual interrelation. Only then will detailed mathematical construction of a definite model be possible.

Let us, therefore, compare some facts relating to stars of the two above types, which seem essential to us when endeavouring to find out the nature of their variability.

T Tauri type stars are of interest since they are found in groups forming associations. These associations were named T-associations by us. It was found later on that this name was quite correct. It was discovered that between the T-associations and the usual O-associations there exists a definite connexion which manifests itself in the fact that some O-associations contain a considerable number of T Tauri variables and are, therefore, at the same time T-associations. A large number of T Tauri stars are contained, for instance, in the Orion association (shown by the studies of Parenago and Haro), and in the associations of Monoceros I and Perseus II, investigated from that point of view by Herbig.

The presence of T Tauri stars in the Perseus II association, which represents, according to Blaauw, a diverging group of hot stars, deserves particular attention. Since the youth of this group may be considered as established, it is natural to suggest that the T Tauri stars are also young objects.

We shall not discuss here some other arguments in favour of the youth of the T Tauri stars. Let us mention only briefly the most probable alternative suggestion; namely, that the T Tauri stars are ordinary dwarfs which have entered into a dust cloud at random. This was suggested by the fact that the T Tauri stars are actually met very often in dark or bright diffuse nebulae.

In Kholopov's study it was shown for the first time that the members of the T-association in Taurus are distributed in small, comparatively compact groups, and that the density of some of these groups is so large that it exceeds the partial density of dwarfs of corresponding luminosities in the surrounding stellar field. This fact contradicts the suggestion regarding ordinary dwarfs entering into the nebulae at random. Herbig has shown more recently that in the association of Monoceros I we observe still denser groups of T Tauri stars. Thus, the hypothesis concerning dwarfs that have accidentally entered a nebula must be rejected, and it should be accepted that the T Tauri stars forming a given group are of a common origin.

On the other hand, since the majority of T Tauri stars are found in diffuse nebulae, it should be considered that these stars lose their T Tauri properties previous to the dissipation of the nebula, or previous to their leaving the nebula. Since, according to modern ideas, the diffuse nebulae are unstable formations and must have ages of the order of 1–2 millions of years, we conclude that the T Tauri type stars cannot be older than 1–2 millions of years. The observed properties of the T Tauri stars must, therefore, be considered as connected with the internal properties of these young objects.

The representatives of the UV Ceti type known to us are located in the immediate neighbourhood of the Sun because, owing to their low luminosities, it is extremely difficult to observe such stars at greater distances. Therefore, the suggestion that there is a nebula associated with such stars is equivalent to the hypothesis that the Sun is located in a diffuse nebula. If we should admit this hypothesis, then we must assume that the density of that nebula is insignificant or at least so small that it exerts no influence upon the spectral features of the Sun. Any influence upon the faint M-type dwarfs which move rapidly across the nebula must be still less. Therefore, the appearance of bright lines in their spectra and, in particular, of outbursts arising under such influence, are out of the question.

Thus, in the case of the T Tauri type stars, as also in the case of the UV Ceti type stars, their variability and spectral properties are connected with the laws of internal evolution of such objects.

One of the most important features connecting the two classes of variable stars under discussion is the presence of continuous emission in the spectra of both types of stars. According to Joy, continuous emission is observed in UV Ceti during its outbursts, whereas in the T Tauri stars it appears at different stages of their light variation.

Over a year and a half ago, when we arrived for the first time at the conclusion that the causes of the continuous emission in the T Tauri and in the UV Ceti stars are similar, the results of Haro and his collaborators on the rapid variables in the Orion nebula and in other associations were not yet known to us. The discovery of rapid variables in T-associations by the Mexican astronomers appeared to link the UV Ceti and T Tauri stars, and showed that these two classes of variable stars, as well as the rapid variables, belong to the same large family of variable dwarfs. The most important property of this family as a whole is the appearance, from time to time, of continuous emission in their spectra. The physical interpretation of the processes going on in the atmospheres of these variables requires, therefore, an understanding of the causes and of the nature of the phenomenon of continuous emission.

Continuous emission manifests itself in various stars in quite different ways with respect to duration and intensity. Therefore, it becomes possible to exclude quite a number of hypotheses on the nature of the continuous emission and to approach, in this way, an understanding of the character of this phenomenon on the basis of known empirical data. The fact that extremely intense continuous emission is observed during outbursts of the UV Ceti type stars makes one assume that the increase of brightness is in these cases mainly caused by the continuous emission. Therefore, the problem of the cause of the increase in brightness is, at least sometimes, coincident with the problem of the continuous emission.

When the increase in brightness is caused by thermal radiation, it must be a consequence of either an increase in temperature, or an increase in the radius of a star.

Cases are known, however, when the increase in brightness occurs within a few seconds. Thus, during the outburst of UV Ceti on 24 October 1952, the brightness of the binary containing this variable increased more than 1·6 magnitudes within 7 sec. This means that the luminosity increased more than 4-fold. If the increase in brightness were to be explained by a change in area, it would signify that the radius of the star had become at least doubled in the course of 7 sec. This would require an expansion of the surface layer with a velocity of 50,000 km./sec. or more. This velocity is totally excluded since the simultaneously observed bright lines do not show any appreciable Doppler shift. It must be assumed, then, that the continuous emission is connected with a considerable increase in the temperature of the outer layers. The possibility that some increase may take place in the temperature of the stellar atmosphere as a result of the appearance of continuous emission cannot be denied. However, it is not this possibility that we are discussing here, but rather the possibility of the appearance of continuous emission itself as a consequence of an increase of temperature of the external layers.

Two possibilities require consideration in this connexion:

(1) The increase of temperature may be caused by an increase in the radiation flux, due to a change in the internal parts of the star. In this case an increase of temperature must take place not only in the photosphere, but also in the deeper parts. But then a decrease of the radiation flux might be possible only after these inner parts have cooled. This would require at least some hours because of the properties of the transport of radiation in the outer layers. However, the entire outburst of UV Ceti on 24 October 1952, including the descending branch of the light curve, took place in only 2 minutes. We must, therefore, reject the hypothesis regarding the heating of the atmosphere as a result of the increase of the radiation flux from the interior.

(2) The increase in the temperature of the external layers may occur as a consequence of the liberation of energy in the same layers. There are two possibilities in this case: (a) thermal emission in the external layers may result at the expense of the energy of some mechanical motion propagated from the internal to the outer parts, or (b) it results at the expense of some other sources of energy. If variant (a) is correct, we must expect something like a blast, embracing either the star as a whole or some of its regions. In this case, the phenomenon should always be of short duration; i.e. the transformation of the blast energy into thermal energy in the external layers and, consequently, the liberation of the continuous emission, should last an extremely short time. The phenomenon of continuous emission has, however, the following property: while in the case of outbursts of the UV Ceti stars the emission is observed for extremely short intervals of time, in some T Tauri stars, DD Tauri and BD + 67° 922 in particular, it is observed to last over a period of years. Hence variant (a) must be rejected, and it must be supposed that continuous emission is liberated at the expense of some other supply of energy available in the atmosphere of the star. Since the total sum of thermal and other types of energy present at a given moment in those outer layers of the atmosphere where continuous emission originates is extremely small, we must admit that the sources of

energy of such emission are brought from the internal parts and only then are liberated in the external layers.

This possibility is apparently the only one which does not directly contradict observations. It is quite natural that if a rapid or slow liberation of some sources of energy (unknown to us at the present) is going on in the outermost layers of the atmosphere, then, owing to the transparency of these layers for continuous radiation, the liberated energy may be emitted without being transformed to any large extent into heat. It would exert, therefore, no major influence upon the mean kinetic energy of the particles in the corresponding layers of the atmosphere. The additional radiation liberated must therefore be of a quite different nature from thermal; i.e. it must be non-thermal radiation.

It must be asked what are the possible sources of energy that could be transferred from the internal to the external layers to give rise to continuous emission. Since we observe in some stars prolonged continuous emission, whose intensity is similar to that of the total thermal radiation of the star, it is natural to assume that the nature of such sources is similar to that of the internal sources of stellar energy: i.e. that these sources are connected with some nuclear processes. It is difficult to say anything definite concerning such nuclear processes at present. It is most probable that these processes of atomic decay are going on, however, not in microscopic atomic nuclei of the usual type, but in nuclear formations on a macroscopic scale, i.e. in such objects which are not yet known to us.

The picture of the changes going on in T Tauri stars is usually more complicated than in UV Ceti stars. Along with the phenomenon of continuous emission of variable intensity, there are also observed variations of colour temperature which cause changes in the thermal radiation. Besides, variations of the intensity of the emission lines are also superimposed upon the variations of the two other kinds.

To explain such a complex of phenomena it must be admitted, as was done in our paper published in the *Communications of the Burakan Observatory*, No. 13, that the liberation of the energy which comes from the internal parts of the star may take place in various layers of its outer envelope. If the energy is liberated below the photospheric layers, we shall observe an additional amount of thermal radiation passing through the photosphere, and comparatively slow light-variations. If the energy is liberated above the photospheric layers, we must observe an increase of the continuous emission and sharp light-variations. It is of interest that the intensity of the continuous emission of the T Tauri stars sometimes increases quickly, although the duration of the maximum may be of considerable length.

In the intermediate case, when energy is liberated in the photospheric layers themselves, we must expect to observe both an increase in the thermal radiation as well as in the continuous emission, together with superimposed absorption lines. In this case, obviously, it will be more difficult to distinguish the continuous emission from the thermal one. Let us finally point out that the bright line spectrum will also change, depending on the depth of the phenomenon.

I shall not discuss here in detail the relation of the phenomenon of continuous emission to the radiation of the comet-shaped nebulae. Data connected with it were communicated at the 1954 Liège Symposium. I should like only to emphasize that in some cases a considerable amount of the radiation of the comet-shaped nebula may be explained by a reflexion of the light of the variable star, but in other cases the reflected light of the variable plays almost no part, and the radiation of the nebula should be ascribed to continuous emission, originating from the direct liberation of energy in the nebula itself.

It seems to us that for a clearer understanding of the processes going on in T Tauri stars, it is extremely important to study those cases in which one aspect or the other of these processes is most pronounced. In other words, a more detailed study of some sub-types of variables of this class is very promising.

Let us discuss here in more detail four species of the above objects. We do not try to classify them, but choose these species only for the sake of emphasizing the necessity of a detailed study of such objects in which characteristic features of the phenomenon of continuous emission are shown most clearly.

First species. Here belong the T Tauri type objects with particularly prolonged and intense continuous emission. Owing to the peculiar distribution of the intensity of continuous emission with frequency, the ultraviolet regions of their spectra are extremely intense.

The most typical representatives of the above species are DD Tauri, studied by Struve and Swings, and LHα 61, discovered by Herbig in his investigation of the association in the neighbourhood of S Monocerotis.

It should be mentioned that the above stars have two more properties in common. The first is the presence of Balmer emission lines out to a high quantum number. The second property is their connexion with comet-shaped nebulae whose brightness greatly exceeds the maximum brightness which would be observed if the nebulae should only reflect the light of the stars.

As was shown by Haro, there are some blue objects among the variable stars of the Orion Nebula. The blue colour of these objects is doubtlessly caused by the distribution of energy in the spectrum of continuous emission. Haro found that in such cases the intensity of the bright Hα line usually is extremely high. Apparently these variables are, according to these properties, closely related to DD Tauri and LHα 61.

The comparatively small number of stars of this species and the particularly intense appearance in them of phenomena typical of the T Tauri stars, testify to the shortness of this stage of their evolution. It is, probably, the earliest stage of the existence of the T Tauri type stars.

Second species. These are the Herbig-Haro Objects. They consist of faint stars surrounded by gaseous nebulae of extremely small diameters. The spectra of these nebulae contain forbidden bright lines of a low degree of ionization. The absolute magnitude of the central stars of these Objects is about $+9^m$, about equal to that of DD Tauri. In spite of their low absolute magnitude, the nuclei of these objects are blue stars, according to Haro. It is natural to suggest in this case also, that the blue colour is not caused by high temperature but by continuous emission. It is of interest that in spite of the extreme rarity of these Objects, three of them located in the Orion association are distributed in the form of a short chain, 5 minutes of arc long. This cannot be a random coincidence and is evidence in favour of the extreme youth of these objects. It would be very desirable to find similar objects in other associations.

Third species. These are the rapid variables discovered by Haro and his collaborators in the Orion Nebula and in Taurus. The absolute magnitudes of these rapid variables are of the same order as the absolute magnitudes of other T Tauri stars, and differ from those of the UV Ceti stars. This species occupies, therefore, an intermediate position between the UV Ceti stars and the ordinary T Tauri stars, filling the intervening gap. It is interesting that the emission lines of these objects are not strong. In the spectra of other objects the continuous emission is accompanied by bright lines. It is, therefore, very important to establish whether in this case the increase of the brightness is connected with the increasing continuous emission, or whether it is caused by the increase of thermal radiation.

Objects belonging to this species, as well as the UV Ceti stars, show a rapid rate of liberation of energy. But it should not, however, be supposed that in other cases, when the phenomenon of continuous emission is prolonged, that the process of liberation of energy is also prolonged. It is not excluded that between the liberation of energy from its sources and its transformation into light quanta of the continuous spectrum there exists one more stage, the duration of which may be quite different in various cases.

Fourth species. This species has at present only one representative, but an

183

extremely interesting one, namely the variable star BD+67° 922 (=AG Draconis). Along with rather intense continuous emission, the following phenomena are typical for this star:

(1) The presence of extremely intense bright H lines and particularly of the bright line of λ 4686 of ionized helium.

(2) Its membership in the spherical component of our Galaxy, which is apparent both from its high galactic latitude (41°) and also from its radial velocity (about -140 km./sec.). The galactic longitude of this star is almost equal to that of the solar apex in respect to the stars of high velocity. The sign of the observed radial velocity is therefore explained. It is of interest that the radial velocity of BD+67° 922 almost coincides with that of the high-velocity long-period variable R Draconis (the period of which is 245 days), located in its vicinity.

It is seen from this example that in the spherical component of our Galaxy there are also T Tauri type stars, the most important difference between +67° 922 and the ordinary T Tauri stars met in associations being the presence of high-excitation He II lines.

As was mentioned above, the liberation of large amounts of energy in the outer parts of the T Tauri type stars and in the outermost layers of their atmospheres may be considered as the physical cause of the processes going on in these stars. It is likely that such liberation of energy is connected with nuclear processes. However, these processes are, according to their nature, altogether different from the already known processes of liberation of nuclear energy and particularly from the thermo-nuclear reactions. The fact that such liberation occurs in the form of explosions suggests a transfer of matter that is in a state of nuclear instability from the internal parts to the external layers. On the other hand, since these phenomena are observed in young stars it may be that the material which is being brought up from the interior contains pre-stellar matter of high density. It may represent matter that is in an altogether peculiar state, thus far unknown to us.

This point of view naturally meets the objection that the problem has here been ascribed to unknown physical processes, whereas not all the possibilities of explanation involving known physical laws have been exhausted.

It should, however, be stated that in studying a phenomenon, no matter how much it has been investigated, we can never be certain that all possible explanations of that phenomenon on the grounds of the existing laws of physics have been tried. Despite this, at some stage in the study, we have to consider that the phenomenon under investigation, which cannot be explained on the basis of known laws of theoretical physics, is a manifestation of some other laws as yet unknown to us. Such a supposition may be incorrect or correct, but unless such suggestions are used to explain some unexpected results of physical experiments and astrophysical observations, no progress in the discovery of new, more profound properties of matter will be made. It should be decided whether the facts relating to the T Tauri type stars really require a similar suggestion. It seems to us that if not only the data involving the spectroscopy and photometry of these stars are taken into account, but also the facts involving the association of these stars in groups of young objects, then a large number of arguments in favour of our suggestion will be obtained.

Some astronomers may disagree with us and may attempt to solve the problem on the basis of known properties of matter. We think, however, that this approach will be unsuccessful. Consequently the rapidly growing knowledge of the T Tauri stars permits us to pass on to the study of the laws of physical processes of a new type, which govern a number of phenomena taking place in these stars.

As an addendum to PROF. AMBARTSUMIAN's paper, MRS MASEVICH presented for him a report on some recent spectrophotometric observations of BD + 67° 922 that were made by Mirsoian with an objective prism and the 8-inch Schmidt camera of the Burakan Observatory.

Spectrograms were obtained in January–February and again in April–May 1955. Both sets of plates revealed continuous emission in the spectrum of BD + 67° 922, but there was a marked difference between them. In January–February, the continuous emission was strong and could be observed from the ultra-violet up to λ 4200. At the second series of observations, the continuous emission was weaker, and extended only up to λ 3900. In the green region of the spectrum, where normal thermal radiation prevails, no change in the brightness was observed. However, in the ultra-violet the change was as much as one magnitude between the two dates.

In these observations of BD + 67° 922, the change in luminosity is apparently caused only by a change in the intensity of the continuous emission. This is regarded as strongly supporting the position taken by Ambartsumian in the foregoing paper, namely that the presence of continuous emission is connected with the existence of a non-thermal process of energy liberation in the stellar atmosphere, above the reversing layer. The operation of this process apparently does not depend upon conditions in the photospheric layers.

The spectral energy distribution of the continuous emission cannot be interpreted, according to Mirsoian, in terms of known physical mechanisms. This conclusion is in good agreement with the very interesting results obtained recently by Herbig and Haro.

Finally, it should be mentioned that the hydrogen emission lines in the spectrum of BD + 67° 922 vary in strength together with the continuous emission. This means that there probably exists a similarity between BD + 67° 922 and the blue T Tauri type stars observed by Herbig in the Orion and Monoceros nebulae.

23. POSSIBLE SOURCES OF INSTABILITY
IN STARS

P. LEDOUX
Institut d'Astrophysique, Cointe-Sclessin, Belgium

(1) INTRODUCTION

The most immediate purpose of the study of stellar stability is to discover the sources of incipient instability which must be responsible for the observed variability of a great number of stars.

I will only discuss here the stability of the star as a whole. Local instabilities such as buoyancy due to a super-adiabatic gradient or peculiar magnetic fields[1] will be considered (only) in so far as they might have an influence on the general stability of the star.

Compared to the classical, mainly mechanical questions of stability, a significant difference is that the thermodynamical factors are here of primary importance. Up to now, two lines of attack have been considered. One which has not received a great deal of attention[2] endeavours to formulate a principle of minimum in analogy to the principle of the minimum of potential energy for mechanical systems. In this connexion, and since all considered systems are open, one may wonder whether the principle of the minimum rate of entropy production and the considerable amount of work done in that field during the last few years[3] might not find interesting applications in some aspects of the problem.

The other method more fully worked out by Jeans, Eddington, Rosseland, Cowling and others is based on the theory of infinitesimal perturbations and has led to the recognition of three main types of instability:

(a) Secular stability, which has been studied mainly for special types of perturbations corresponding to homologous transformations[4] and which is of interest for very slow changes such as have been usually associated with the notion of stellar evolution.

(b) Dynamical stability, which is realized if the star, subjected to a small adiabatic perturbation varying with time as $e^{i\sigma t}$, can oscillate with a finite real frequency σ. Instability here is connected with the occurrence of imaginary frequencies leading to 'explosions'.

(c) Vibrational stability, which characterizes the variation with time of the amplitude of the oscillation when deviations from isentropic motion are taken into account.

Here, we are primarily interested in the last two types of instability. Do the observations provide any clue as to the type of instability present in different classes of non-stable stars? Table 1 contains, for some of these stars, three of the characteristics which seem most fundamental in that respect: the period or time separation $\tau_{\text{obs.}}$ in days between similar phases (when the phenomenon repeats itself), reduced by the homology factor $\sqrt{\bar{\rho}/\bar{\rho}\odot}$ where $\bar{\rho}$ is the mean density; the ratio $\Delta L/L$, where ΔL is the maximum deviation from the average (or normal) luminosity L; and the ratio $E/(L_{\text{min.}} \times \tau_{\text{obs.}})$, where E is the total energy emitted during the outburst.

Nearly all figures given are subject to considerable uncertainties. In particular, the computation of $\bar{\rho}$ rests upon the use of the mass-luminosity relation and the relation between radius, luminosity and spectral type, both of which might show systematic effects when going from population I to population II. However, they are probably sufficient for a qualitative discussion.

Table 1. *Some Fundamental Characteristics of Non-Stable Stars*[5]

Class	$\tau_{\text{obs.}} \sqrt{\bar{\rho}/\bar{\rho}\odot}$	$\Delta L/L$	$\dfrac{E}{L_{\text{min.}} \times \tau_{\text{obs.}}}$
Classical cepheids	0·04		1 to $\frac{1}{2}$
Red semi-regular variables	$\simeq$0·04		
Long period variables	$\simeq$0·07		
RR Lyrae	0·06	$\frac{1}{2}$	1 to $\frac{1}{2}$
W Virginis stars	0·07–0·15		
Short period variables (in clusters)	$\simeq$0·15		
RV Tauri stars	$\simeq$0·15–0·30	5–10	1 to $\frac{1}{2}$
SX Herculis stars	$\simeq$0·30–0·40		
β Canis Majoris stars	0·04–0·09	Small	
SX Phoenicis	0·08		
Flare stars	10	5–100	Small $\simeq$0·01
T Tauri stars	?	10–20	
RW Aurigae stars	?		
Z Camelopardalis stars	Large	30–50	
U Geminorum stars	500		3 to 5
Recurrent novae	10^4–10^5	Very large	1 to $\frac{1}{2}$
Classical novae	Very large		1 to $\frac{1}{2}$?
Super-novae	∞ or extremely large	Extremely large	

The figures in the second column of Table 1 should be compared to the theoretical periods of oscillation. As very complicated modes are unlikely, we can use for this purpose the periods of the fundamental mode of radial pulsation which are given for different models in Table 2.

On this basis, one would be tempted to consider roughly two main groups. In the first one (regular variables and some semi-regular or irregular variables), theoretical and observed periods are of the same order of magnitude and $\Delta L/L$ is of the order of unity. The problem here is essentially that of explaining the appearance and the maintenance of the oscillations: that is, it is a problem of *vibrational instability*.

Table 2. *Theoretical Periods* $\tau_{th}\sqrt{\bar{\rho}/\rho\odot}$ *for Different Models*

	Homogeneous model	Polytrope $n=1\cdot5$ (convective model)	Standard model	Original[6] Epstein model	Modified[7] Epstein model (with large external convection zone)
$\rho_c/\bar{\rho} =$	1	6	54	2×10^6	$1\cdot2\times10^6$
$\tau_{th}\sqrt{\bar{\rho}/\rho\odot} =$	0·1156	0·075	0·039	0·031	0·056

In the second group, the length of the cycle increases extremely rapidly as compared to the theoretical periods which, probably, have no significance for the observed phenomena. At the same time, $\Delta L/L$ becomes very large. It would seem that in those cases, the stars are on the verge of some kind of *dynamical instability* which manifests itself from time to time. On the other hand, if we exclude the flare stars, the last column of Table 1 (which reproduces values quoted by Schatzman[8] in discussing an extension of the Kukarkin-Parenago relation) would rather suggest some kind of underlying unity between all these phenomena. A detailed investigation of this point would be interesting, but we should note that, in these considerations, the phase of minimum luminosity plays a privileged role and this is perhaps not at all justified, especially for the regular variables.

Before discussing the theory, let us recall that the discovery of an instability by the perturbation method does not necessarily mean that the whole star will cease to exist, since the increase of the amplitudes to finite values might very well remove the cause of instability or call in stabilizing factors.

(2) DYNAMICAL INSTABILITY

There is a rather significant difference between radial and non-radial perturbations and we shall discuss them separately.

(a) Radial perturbation

It is well known that in this case the formulation of the problem leads to a linear equation of the Sturm-Liouville type where the square of the frequency σ^2 plays the role of the parameter.

The eigenvalues σ_i^2 ($i = 0, 1, 2, \ldots$) ordered by increasing values correspond to eigenfunctions ξ_i defining the relative displacements $(\delta r/r)_i$ for successive modes with 0, 1, 2, . . . i nodes between the centre and the surface. Thus any instability ($\sigma^2 < 0$) will manifest itself first through the fundamental mode (σ_0, ξ_0), and it is then sufficient to discuss this case. Its frequency is given by[9]

$$\sigma_0{}^2 = -\frac{\int_0^R 4\pi\xi_0 r^3 \frac{d}{dr}[(3\Gamma_1 - 4)P]\,dr}{\int_0^R \xi_0\, 4\pi\rho r^4\, dr} = \frac{\int_0^R (3\Gamma_1 - 4)\,P\left(-\frac{\delta\rho}{\rho}\right)_0 4\pi r^2\, dr}{\int_0^R \xi_0\, 4\pi\rho r^4\, dr}, \quad (1)$$

where ξ_0 and $-(\delta\rho/\rho)_0$ are everywhere positive and Γ_1 is a generalized ratio of specific heats for a mixture of radiation and ionized gas, the nuclei of which might be taking part in thermo-nuclear reactions or might even have reached a state of nuclear equilibrium. If Γ_1 is constant, one finds immediately the well-known result that the condition for instability is

$$\Gamma_1 < \tfrac{4}{3}. \quad (2)$$

Of course, strictly speaking, Γ_1 will never be a constant, as the ionization of an electronic shell of an abundant element can lower its value appreciably and even render it smaller than $\tfrac{4}{3}$ in the region of the star where it takes place.

However, with the accepted predominance of H and He and for normal stars, these layers of low Γ_1 are rather narrow and only occur fairly close to the surface of the star, where the product $P(\delta\rho/\rho)_0$ is small so that they affect very little the appropriate mean $\overline{\Gamma}_1$ (cf. equation (1)), which in this case should replace Γ_1 in equation (2).

A negligible abundance of He and H would be more favourable especially for large masses, where the pressure of radiation would tend to decrease Γ_1 everywhere. However, a detailed discussion of the standard model[10] has failed to reveal any physically significant case of instability for normal dimensions.

Of course, the model chosen might influence the result somewhat since the region of greater weight for $(3\Gamma_1 - 4)$ will displace itself according to the central condensation. However, a rapid comparison between Epstein's model and the standard model did not disclose any important differences. One may then conclude that the chances of dynamical instability towards purely radial perturbations of any aggregate of ordinary matter of stellar dimensions are extremely small.

However, Biermann and Cowling have pointed out[2] that if one goes to a larger and larger radius for a given mass, a configuration devoid of H and He will finally reach a state of dynamical instability for sufficiently large dimensions. A slight admixture of H and He will increase the critical radius and it would be worth while to extend their discussion to the case of large H and He abundances, as it might have interesting cosmological applications in problems such as the formation of stars by condensation of interstellar clouds.

Up to now, we have only considered the effects of ionization on Γ_1, but, of course, the influence of nuclear reactions on these problems could also be discussed through their effects on Γ_1. In fact, thermo-nuclear reactions which lead to a rate of generation of energy ϵ, directly proportional to some powers of the temperature T and density ρ, would add an imaginary part to Γ_1 that would affect the vibrational stability of the star, so that this type of reaction is of no interest here.

On the other hand, in the case of a real nuclear equilibrium, ϵ is proportional to the time derivative of T and ρ, and nuclear reactions contribute a real term to Γ_1. If T and ρ are large enough so that an equilibrium is established between nuclei and elementary particles such that any further increase in T or ρ leads to further dissociation of complex nuclei, Γ_1 could be reduced to a value close to 1 in a large part of the star and this could lead to a violent instability.

For instance, in Hoyle's theory [11] of the formation of heavy elements in stars, the phase of collapse under gravity and the reversal of this into a phase of explosive expansion by means of the centrifugal force is a manifestation of dynamical instability due to this cause.

According to Hoyle, this could explain the origin of super-novae. This very interesting case should be studied anew carefully from the point of view of dynamical instability using quantitative arguments.

(b) Non-radial perturbation

The problem of non-radial perturbations is more complicated, but it presents an interesting possibility of interaction between what we call here dynamical stability and the stability towards convection, which is insured by Schwarzschild's criterion:

$$A = \frac{1}{\rho}\frac{d\rho}{dr} - \frac{1}{\gamma P}\frac{dP}{dr} < 0. \tag{3}$$

If equation (3) is violated, it can be shown [12] that non-radial perturbations of sufficiently small wave-lengths will be unstable. On the other hand, the stability of purely radial oscillations (infinite horizontal wave-length) is not affected by the sign of A. What does happen in the interesting case of intermediate wave-lengths? The problem is difficult, but there are a few cases where it can be discussed more or less completely. Let us consider first the case of the homogeneous compressible model which, as Pekeris has shown [13], is amenable to a complete analytical treatment and which is highly unstable towards convection:

$$A = -\frac{1}{\gamma P}\frac{dP}{dr} > 0. \tag{4}$$

Considering non-radial perturbations represented by series of terms of the form

$$f(r)\, P_s^m(\cos\theta)\, e^{\pm im\phi} e^{i\sigma t}, \tag{5}$$

s and m being integers, and $-s < m < s$, Pekeris has shown that as soon as one departs from purely radial perturbations ($s = 0$), unstable modes appear. Table 3 summarizes the numerical results obtained by Mme Sauvenier-Goffin [14] for the first few modes. In this Table, s refers to the degree of the spherical harmonics and k to the number of modes of $\delta\rho$ along the radius.

Table 3. *Values of* $\beta = \dfrac{3\sigma^2}{4\pi G\rho}$ *for* $\Gamma_1 = 5/3$

s		$k=0$	$k=1$	$k=2$	$k\to\infty$
0	f	1	12·666	31	$\to\infty$
2	p	8·39	26·23	51·12	$\to\infty$
	g	−0·73	−0·225	−0·117	$\to0$
3	p	12·0	33·03	61·20	$\to\infty$
	g	−1·0	−0·363	−0·196	$\to0$
4	p	15·61	39·84	71·28	$\to\infty$
	g	−1·281	−0·502	−0·2806	$\to0$
0	f	1	12·666	31	$\to\infty$

One notices that for each value of s ($s \neq 0$) and k there are two modes, one with an important vertical displacement δr corresponding to the largest value of σ^2 and another for which the displacement is mainly horizontal; these were called p- and g-oscillations by Cowling. It is through the g-modes that the instability manifests itself, and it increases with s and decreases when k increases: in other words, the most unstable perturbation is that with the smallest horizontal wave-length $\left(\propto \dfrac{1}{s}\right)$, but the largest vertical wave-length $\left(\propto \dfrac{1}{k}\right)$. One may expect that this will remain true generally [12] and a recent paper by Skumanich [15] has confirmed it again in the particular case of a polytropic atmosphere. Of course, viscosity and heat conduction might decrease the instability of some of the higher modes or even make them stable [16].

The next simplest case is that of the polytropes discussed by Cowling[17] in neglecting the perturbation of the gravitational potential. Cowling's reasoning can be extended[18] easily to the general case, starting from the basic equations:

$$\frac{d\phi}{dr} + \frac{1}{\Gamma_1 P}\frac{dP}{dr}\phi = \left[\frac{s(s+1)}{\sigma^2} - \frac{r^2\rho}{\Gamma_1 P}\right]y - \frac{s(s+1)U'}{\sigma^2},\tag{6}$$

$$\frac{dy}{dr} + yA = \frac{1}{\gamma^2}(\sigma^2 + Ag)\phi + \frac{\partial U'}{\partial r},\tag{7}$$

where $\phi = r^2\delta r$, $y = P'/\rho$, P' and U' being the eulerian perturbations of P and U, and $g = -\frac{1}{\rho}\frac{dP}{dr}$. The elimination of U' is difficult, however, and leads to a fourth order differential equation with very complicated coefficients. If, following Cowling, one neglects U', and introduces variables

$$v = r^2\delta r\, P^{1/\Gamma_1} \quad \text{and} \quad w = y\rho P^{-1/\Gamma_1} = P'P^{-1/\Gamma_1},$$

then equations (6) and (7) can be written:

$$\frac{dv}{dr} = \left(\frac{s(s+1)}{\sigma^2} - \frac{r^2\rho}{\Gamma_1 P}\right)\frac{w}{\rho}P^{2/\Gamma_1},\tag{8}$$

$$\frac{dw}{dr} = \frac{1}{r^2}(\sigma^2 + Ag)vP^{-2/\Gamma_1}\rho.\tag{9}$$

Eliminating w and v successively, one finds

$$\frac{d}{dr}\left[\frac{\rho P^{-2/\Gamma_1}}{\dfrac{s(s+1)}{\sigma^2} - \dfrac{r^2\rho}{\Gamma_1 P}}\frac{dv}{dr}\right] = \frac{1}{r^2}(\sigma^2 + Ag)\rho P^{-2/\Gamma_1}v,\tag{10}$$

$$\frac{d}{dr}\left[\frac{r^2 P^{2/\Gamma_1}}{(\sigma^2 + Ag)\rho}\frac{dw}{dr}\right] = \left[\frac{s(s+1)}{\sigma^2} - \frac{r^2\rho}{\Gamma_1 P}\right]\frac{w}{\rho}P^{2/\Gamma_1}.\tag{11}$$

If σ^2 becomes large (p-modes), and one neglects $s(s+1)/\sigma^2$ in comparison to $r^2\rho/\Gamma_1 P$ and Ag in comparison to σ^2, equation (10) becomes

$$\frac{d}{dr}\left(\frac{\Gamma_1 P^{1-2/\Gamma_1}}{r^2}\frac{dv}{dr}\right) + \frac{\sigma^2}{r^2}\rho P^{-2/\Gamma_1}v = 0,\tag{12}$$

which is of the Sturm-Liouville type and admits a spectrum of positive eigenvalues σ_p^2 increasing with the order of the mode considered.

In the same way, if σ^2 becomes small (g-modes), equation (11) reduces to

$$\frac{d}{dr}\left(\frac{r^2 P^{2/\Gamma_1}}{Ag\rho}\frac{dw}{dr}\right) - \frac{s(s+1)}{\sigma^2}\frac{P^{2/\Gamma_1}}{\rho}w = 0,\tag{13}$$

which is again of the Sturm-Liouville type. If A is everywhere negative (condition (3) for thermal stability satisfied everywhere), (13) admits a spectrum of positive eigenvalues σ_g^2, decreasing with the order of the mode considered. In that case, the g-modes as well as the p-modes are all stable.

If A is positive everywhere (condition (3) for thermal stability violated everywhere), the eigenvalues σ_g^2 of (13) will all be negative, and this time, all the g-modes will be unstable.

Multiplying (13) by w and integrating the first term by parts and noting that the integrated part is zero, one gets

$$\sigma_g^2 = -\frac{\displaystyle\int_0^R s(s+1)\, P^{2/\Gamma_1}\rho^{-1}\, w^2 dr}{\displaystyle\int_0^R \frac{r^2 P^{2/\Gamma_1}}{Ag\rho}\left(\frac{dv}{dr}\right)^2 dr}, \tag{14}$$

confirming the fact that if A is of one sign everywhere, σ^2 is of the opposite sign. If A changes sign, one sees that it is the sign of an appropriate mean value of A which determines the stability of the star.

However, the results obtained from these approximate considerations can only be considered as indications. In particular, the passage from (10) and (11) to (12) and (13) involves some delicate points since, close enough to the centre or the surface, the terms neglected become of the same order as the terms kept.

For instance, one could integrate directly (10) multiplied by v and obtain

$$\sigma^2 \int_0^R \frac{\rho v^2}{r^2 P^{2/\Gamma_1}}\, dr + \int_0^R \frac{gA\rho}{r^2 P^{2/\Gamma_1}}\, v^2 dr + \int_0^R \frac{\rho\left(\frac{dv}{dr}\right)^2 dr}{\left(\frac{s(s+1)}{\sigma^2} - \frac{\rho r^2}{\Gamma_1 P}\right) P^{2/\Gamma_1}} = 0,$$

or using (8),

$$\sigma^2 \int_0^R \frac{\rho v^2}{r^2 P^{2/\Gamma_1}}\, dr + \int_0^R \frac{gA\rho}{r^2 P^{2/\Gamma_1}}\, v^2 dr + \int_0^R \left(\frac{s(s+1)}{\sigma^2} - \frac{r^2\rho}{\Gamma_1 P}\right)\frac{w^2 P^{2/\Gamma_1}}{\rho}\, dr = 0.$$

If σ^2 is large, one then gets

$$\sigma_p^2 \int_0^R \frac{\rho v^2}{r^2 P^{2/\Gamma_1}}\, dr = \int_0^R \frac{r^2\rho}{\Gamma_1 P}\frac{w^2}{\rho}\, P^{2/\Gamma_1} dr - \int_0^R \frac{gA\rho}{r^2 P^{2/\Gamma_1}}\, v^2 dr, \tag{15}$$

which shows that even for the p-modes, the values of A play a certain role.

For the g-modes (σ^2 small), one would get in the same way,

$$\frac{1}{\sigma_g^2}\int_0^R s(s+1)\frac{w^2}{\rho}\, P^{2/\Gamma_1} dr = \int_0^R \frac{r^2\rho}{\Gamma_1 P}\frac{w^2}{\rho}\, P^{2/\Gamma_1} dr - \int_0^R \frac{gA\rho}{r^2 P^{2/\Gamma_1}}\, v^2 dr, \tag{16}$$

which again shows that even for the g-modes, A alone does not really determine the sign of σ^2.

One can easily imagine also that in cases close to instability $(\sigma_g \to \pm 0)$, the neglected terms in U' could have important effects.

Any general progress in this problem would certainly be very useful, but even the detailed study of a few special cases would already be very interesting.

The interest of these possibilities of dynamical instability connected with convective instability arises from the fact that the last type of instability must be very common.

Of course, we usually admit that convective equilibrium will replace radiative equilibrium as soon as condition (3) is violated and this is in reasonable agreement with the result mentioned previously, that the most unstable perturbations are those of small horizontal extent.

However, in a region in convective equilibrium, we know that the actual gradient always remains slightly super-adiabatic so that A remains equal to a small positive quantity ϵ. If this situation prevails in the whole star, it is very likely, from our previous discussion, that there will be unstable g-oscillations, and it would be very interesting to compute the lowest degree s of the harmonic which can become unstable, let us say, for the first mode.

The same type of problem should also be solved for a star comprising a convection zone and a part in radiative equilibrium. It is possible that many minor and erratic changes could be traced back to this type of instability arising in an external convection zone.

Of course, the situation would be much more favourable to this type of dynamical instability, if A could become large and positive in an important external zone as was once proposed by Biermann[19]. In that case, one might expect a violent instability for fairly low harmonics which would lead to the ejection of material in the form of a few separated jets. Indeed, this seems to occur at least in some novae.

(3) VIBRATIONAL INSTABILITY

Here we want to study the influence of the deviations from isentropic motion. We shall neglect friction for the time being and admit that the oscillation takes place through a true equilibrium state in which

$$\epsilon_0 - \frac{1}{\rho_0} \operatorname{div} \vec{F_0} = 0, \tag{17}$$

where ϵ_0 is the rate of generation of energy at equilibrium and $\vec{F_0}$, the flux.

In that case, the theory of perturbations[20] shows that the star will be vibrationally unstable towards the k-mode of oscillation if

$$\int_0^M \left(\frac{\delta T}{T}\right)_k \left(\delta\epsilon - \frac{d\delta L(r)}{dm(r)}\right) dm > 0, \tag{18}$$

where $L(r) = 4\pi r^2 F(r)$.

Since for all nuclear reactions of interest, ϵ increases with ρ and T, $\delta\epsilon$ will be of the same sign as δT and the first term of the integral in (18) will always contribute to the instability. As Eddington was the first to point out, however, phase-delays depending on the ratio of the period $(2\pi/\sigma_k)$ to the mean lives of the nuclei may occur in the different reactions considered and some care must be taken in the evaluation of $\delta\epsilon$. One can always write

$$\delta\epsilon = \epsilon_0 \left(\mu_{\text{eff.}} \frac{\delta\rho}{\rho} + \nu_{\text{eff.}} \frac{\delta T}{T}\right), \tag{19}$$

but $\mu_{\text{eff.}}$ and $\nu_{\text{eff.}}$ may be appreciably different from the exponents in $\epsilon_0 = C\rho^\mu T^\nu$ at equilibrium[21]. However, in the current models for main sequence and giant stars, $\mu_{\text{eff.}} \simeq \mu = 1$ and $\nu_{\text{eff.}} \leqslant 20$.

If the heat capacity of the external layers is small, $\delta L(r)$ remains everywhere in phase with δT and for all ordinary opacity laws, the main term in $d[\delta L(r)]/dm$ is proportional to $d(\delta T/T)/dm$ and in a first approximation

$$\frac{\delta T}{T} \frac{d[\delta L(r)]}{dm} \propto \frac{d\left(\frac{\delta T}{T}\right)^2}{dm}. \tag{20}$$

Thus the second part of the integral in (18) will always be negative and contribute to the stability if $(\delta T/T)$ increases in absolute value with $m(r)$. For the fundamental mode of radial oscillations, this is certainly the case for all these models.

Furthermore, the first part of the integral can be limited to the region where the generation of energy takes place, that is, close to the centre where the amplitudes $\delta\rho/\rho$, $\delta T/T$ are small. The second part gets its main contribution from the external layers where the amplitudes are large. As Cowling[22] was the first to show, this makes ordinary stars extremely stable.

With increasing masses, the increasing pressure of radiation lowers the value of Γ_1 and Γ_3, and this reduces the increase of $(\delta r/r)$ from the centre to the surface and vibrational instability becomes possible if $\nu_{\text{eff.}} \simeq 20$, for masses of the order of 100 $m_\odot$[23].

There has been very little discussion of the vibrational stability towards higher modes of radial oscillations but, except in very exceptional cases[23], it should even become reinforced since for a given value $(\delta r/r)_0$ at the centre, the value $(\delta r/r)_R$ at the surface increases with the order of the mode. Furthermore, the damping due to friction increases rapidly with the order of the mode.

As far as non-radial oscillations are concerned, $\delta p/\rho$ and $\delta T/T$ tend towards zero at the centre, and the tendency to instability due to nuclear reactions would practically disappear altogether in this case. Since the damping due to friction also increases very rapidly here with the degree of the harmonics, it is very probable that the vibrational stability would again be reinforced.

Thus, from now on, we will consider only the fundamental mode of radial pulsation. However, we have seen that for ordinary models, the prospects of vibrational instability are rather poor.

What kind of changes could bring about vibrational instability?

Let us assume first that the calorific capacity of the external layers remains small to avoid any phase difference between δT and $\delta L(r)$. If, furthermore, we keep to models with large central condensations and values of Γ_1 close to $\frac{5}{3}$, it would seem that the only possibility consists in displacing the zone of generation of energy towards the surface. But we are limited by the well-known hydrostatic difficulty of building models with large iso-thermal cores [23]. Furthermore, since ϵ must be sensitive to ρ and T in order to play a role in condition (18), it seems impossible at the same time to have an appreciable fraction of the luminosity L generated in the external part of gaseous stars.

Changes in the opacity law do not seem to be very promising either [23] in ordinary stars.

In cool stars, surface phenomena which have been invoked repeatedly such as the formation of 'veils' or clouds could play a role, but although they undoubtedly could be responsible at least for light variations, it is difficult to find a natural way of adjusting them to the right phases in order to give the type of instability we are looking for as well as oscillations with a period depending upon the mean density of the star.

According to Schatzman [21], at very high densities some nuclear reactions might have large values of $\mu_{\text{eff.}}$ or $\nu_{\text{eff.}}$. But to realize these conditions towards the centre of a normal star, the central condensation would have to be so large that the increase of $(\delta r/r)$ from $r=0$ to $r=R$ would be enormous, and it is not sure that even those high values of $\mu_{\text{eff.}}$ and $\nu_{\text{eff.}}$ would make such a star unstable.

Of course, the best way would be to reduce the ratio $(\delta r/r)_R/(\delta r/r)_0$. But this requires either a change in the model towards smaller central con-densation or an appreciable lowering of Γ_1. Both these occur, for instance, in white dwarfs, and there vibrational instability is difficult to avoid[24] if nuclear reactions take place. But we know that other considerations also tend to exclude them[25]. But for normal gaseous stars, a decrease in Γ_1 seems unlikely and a decrease in central condensation would create new difficulties in explaining the generation of energy at equilibrium.

One might think also of introducing some kind of very strong viscosity in a large external part of the star. Of course this would imply an extra damping proportional to

$$\phi = \tfrac{4}{3}\sigma^2 \mu r^2 \left[\frac{d\left(\frac{\delta r}{r}\right)}{dr}\right]^2 , \tag{21}$$

where μ is the coefficient of dynamical viscosity, but at the same time it would tend to make $(\delta r/r)$ a constant in that part. Thus, since the effect of viscosity would decrease the variation of $(\delta r/r)$, the damping due to viscosity itself would decrease as well as that due to the flux, the destabilizing effects of the generation of energy would increase, and one might possibly reach a state of balance.

Of course ordinary viscosity (molecular or radiative) is much too small. Even if one corrects Persico's[26] result by taking into account a large abundance of hydrogen which gives a value of μ that is 200 to 300 times larger, the damping time for the fundamental mode due to viscosity alone remains much too large (of the order of 10^{11} to 10^{13} years, depending on the model). But turbulent viscosity taken proportional to $\bar{\rho c l}$ might be 10^{11} to 10^{12} times larger than ordinary viscosity, and if it would act in a sufficiently large region, a reconsideration of the problem might be worth while.

Finally, one could try to introduce a phase lag between δL and δT through the effects of non-adiabatic processes in the external layers, which must now have an appreciable heat capacity. In that case, the contribution to the integral in (18) of the term

$$\left(\frac{\delta T}{T}\right) \cdot \frac{d[\delta L(r)]}{dm} ,$$

where $\delta T/T$ is only the adiabatic part of the temperature variation, can be made as small as one wants provided the phase lag between $(\delta T/T)_{\text{ad.}}$ and δL is properly adjusted.

Eddington[27] thought that the convection zone of hydrogen, which is capable of storing energy during the contraction and of liberating it during the expansion, could provide the necessary heat capacity. This point does not seem to have been definitely settled. Recently Zhevakin[28] has considered the effects of the ionization zone of helium and he believes that this effect would raise the capacity sufficiently.

In Eddington's theory, the phase of the displacement δr is affected very little so that the wave remains mainly a standing wave. However, the case, first considered by Schwarzschild[29], where the wave takes a progressive character in the external layers is also very interesting. However, there does not seem to be any discussion of vibrational stability for such a case. For the simple adiabatic progressive wave discussed by Schwarzschild in his first papers on the subject, it is possible to show that the instability with respect to the case of a standing wave is not increased. But it would be interesting to consider more general types of progressive waves.

(4) CONCLUSIONS

Our review of the more classical aspects of the theory did not provide us with many possibilities for instability dynamical or vibrational, and the few cases left open will require probably a good deal of critical study before definite conclusions can be reached.

We have neglected some factors such as rotation or magnetic fields, for instance, which might be important at least in some cases [1, 30], but I doubt that they could contribute to a general instability for normal stars. Schatzman[21, 31] has introduced an interesting idea, according to which a nuclear explosion could be started by an oscillation which becomes vibrationally unstable. But apart from difficulties in following the effects of this explosion, the main problem is still to discover the source of the vibrational instability. It is true, however, that if this idea is used to try to explain a nova outburst, the normal state of the star must be rather peculiar, approaching the white-dwarf stage where vibrational instability might be more common.

One should note also that the perturbation method used here does not necessarily cover all possible cases of instability or periodic variations. In this last respect, non-linear oscillations have received little attention, but one must admit that up to now, the evidence for physical factors capable of maintaining such oscillations seems to be lacking[32].

On the other hand, the purely hydrostatic approach to the problem of the internal structure of stars has revealed that evolutionary sequences of models can, for instance, lead to critical situations such as a maximum convective or isothermal core or very critical conditions for the existence of radiative or convective equilibrium in a part of the star. It would be interesting in these cases to replace the hydrostatic approach by a dynamical one taking into account non-linear terms. This might reveal interesting new possibilities, such as non-static stellar models oscillating periodically around a fictitious position of equilibrium.

REFERENCES

[1] Cf. S. Chandrasekhar, *Proc. Roy. Soc.* A, **225**, 173 (1954); *Phil. Mag.*, ser. 7, **45**, 1177 (1954); E. M. Parker, *Ap.J.* **121**, 491 (1955); E. Jensen, *Ann. d'Ap.* **18**, 127 (1955); M. Kruskal and M. Schwarzschild, *Proc. Roy. Soc.* A, **223**, 348 (1954).
[2] L. Biermann, *Zs. f. Ap.* **16**, 29 (1938); L. Biermann and T. G. Cowling, *Zs. f. Ap.* **19**, 1 (1939); R. C. Tolman, *Ap.J.* **90**, 541, 568 (1939).
[3] Cf. I. Prigogine, *Etude thermodynamique des phénomènes irréversibles*, Desoer, Liège, 1947.
[4] Cf. for a more general attack of this problem: L. H. Thomas, *M.N.* **91**, 122 (1930), and an interesting discussion in: S. Rosseland, *The Pulsation Theory of Variable Stars*, Oxford, 1949, §§5.12 to 5.14.
[5] Table 1 is based mainly on data taken from C. Payne-Gaposchkin, *Variable Stars and Galactic Structure*, University of London, 1954.
[6] L. Epstein, *Ap.J.* **112**, 6 (1950).
[7] P. Ledoux, J. Bierlaire and R. Simon, *Ann. d'Ap.* **18**, 65 (1955).
[8] E. Schatzman, *Les Principes fondamentaux de classification stellaire*, C.N.R.S., Paris, 1955, p. 176.
[9] P. Ledoux, 'Contribution à l'étude de la structure interne des étoiles et de leur stabilité', *Mém. Soc. Roy. Sci. Liège*, 4° série, **9**, 235 (1949).
[10] Cf. reference [9], chapter IV.
[11] F. Hoyle, *M.N.* **106**, 343 (1946).
[12] For a general discussion of this point cf. reference [9], the beginning of chapter III.
[13] C. L. Pekeris, *Ap.J.* **88**, 189 (1939).
[14] E. Sauvenier-Goffin, *Bull. Soc. Roy. Sci. Liège*, **20**, 20 (1951).
[15] A. Skumanich, *Ap.J.* **121**, 408 (1955).
[16] S. Chandrasekhar, *Phil. Mag.*, ser. 7, **43**, 1317 (1952) and **44**, 233 (1953).
[17] T. G. Cowling, *M.N.* **101**, 367 (1941).
[18] P. Ledoux, *III° Congrès National des Sciences*, Bruxelles, 1950, **2**, p. 133.
[19] L. Biermann, *Zs. f. Ap.* **18**, 344 (1939).
[20] Cf. S. Rosseland: *The Pulsation Theory of Variable Stars*, Oxford, 1949, chapter V.
[21] Cf. reference [20], chapter V, §5.3: P. Ledoux and E. Sauvenier-Goffin, *Ap.J.* **111**, 611 (1950); E. Schatzman, *Ann. d'Ap.* **14**, 305 (1951) and *Les processus nucléaires dans les astres*, *Mém. Soc. Roy. Sci. Liège*, **14**, 163 (1954); D. A. Franck-Kamenetsky, *Comptes rendus de l'Acad. d. Sci. de l'U.R.S.S.* **77**, 385 (1951).
[22] T. G. Cowling, *M.N.* **96**, 42 (1936).
[23] P. Ledoux, *Ap.J.* **94**, 537 (1941).
[24] P. Ledoux and E. Sauvenier-Goffin, *Ap.J.* **111**, 611 (1949).
[25] L. Mestel, *M.N.* **112**, 583 (1952).
[26] E. Persico, *M.N.* **86**, 93 (1926).
[27] A. S. Eddington, *M.N.* **101**, 182 (1941); **102**, 154 (1942).
[28] S. A. Zhevakin, *A.J. U.S.S.R.* **32**, 161 (1953).
[29] M. Schwarzschild, *Zs. f. Ap.* **15**, 14 (1938); *Harv. Circ.* Nos. 429, 431 (1938).
[30] S. Chandrasekhar and D. H. Nelson Limber, *Ap.J.* **119**, 10 (1954).
[31] E. Schatzman, *Ann. d'Ap.* **14**, 294 (1951); **17**, 152 (1954).
[32] S. A. Zhevakin, *Comptes rendus de l'Acad. d. Sci. de l'U.R.S.S.* **62**, 191 (1948); P. Ledoux, *Bull. Ac. Roy. Belgique, classe des Sci.*, 5° ser., **38**, 352 (1952); W. S. Krogdahl, *Ap.J.* **122**, 43 (1955).

DISCUSSION

In the discussion of the papers by Prof. Ambartsumian and Dr Ledoux, Dr E. Schatzman expressed himself as being much impressed by the arguments of Ambartsumian concerning the origin of the non-thermal radiation in the stars. He regarded it as very probable that the explanation of this phenomenon in the UV Ceti and T Tauri stars, and in the variable cometary nebulae, lay in the domain of nuclear processes. However, Schatzman wished to propose a different explanation from that of Ambartsumian for this nuclear phenomenon. He stated that it seems possible that hydromagnetic processes would be able to communicate considerable velocities to atoms in the outer layers of certain stars. During this process, nuclear reactions would take place, and Schatzman thanked Mrs Burbidge for having drawn his attention to the resonance reaction $Ne^{20} + H^1 \rightarrow Ne^{21} + e^+$, and to the subsequent formation of neutrons from the reaction $Ne^{21} + He^4 \rightarrow Mg^{24} + n$. These neutrons would lead to the formation of radioactive elements, which are carried out to produce the non-thermal radiation of the variable comet-like nebulae.

Regarding the paper of Dr Ledoux, Schatzman remarked that the value of $E/L_{min.} \times \tau_{obs.}$ for the UV Ceti stars, being several orders of magnitude smaller than for the other stars of Table 1, indicates that the flare-star phenomenon is of an entirely different nature from the processes operating in the other variables. It therefore seems necessary, according to Schatzman, to take magnetic phenomena into account in the study of the instability of the outer layers of such stars.

Mrs E. M. Burbidge stated that she does not believe that the production of neutrons by the reaction mentioned by Dr Schatzman, which may occur in stellar interiors, would be important on the stellar surface. The production of neutrons by reactions between protons and light nuclei might occur, but recent work has suggested that it will do so only if excited regions appearing through violent electro-magnetic disturbance on the stellar surface have much higher kinetic temperatures than those needed to explain the brightness variations in flare stars. If such regions on flare stars achieve equivalent temperatures of 50,000° or less, as would explain the flaring characteristics, Mrs Burbidge felt that nuclear reactions would be of very small importance.

Printed in the United States
By Bookmasters